Non-Destructive Testing Techniques

Non-Destructive Testing Techniques

RAVI PRAKASH

Dean, Research & Consultancy Division
Birla Institute of Technology and Science
Pilani, India

New Age Science Limited
The Control Centre, 11 A Little Mount Sion
Tunbridge Wells, Kent TN1 1YS, UK
www.newagescience.co.uk • e-mail: info@newagescience.co.uk

Copyright © 2009 by New Age Science Limited

The Control Centre, 11 A Little Mount Sion, Tunbridge Wells, Kent TN1 1YS, UK

www.newagescience.co.uk • e-mail: info@newagescience.co.uk

Tel: +44(0) 1892 55 7767, Fax: +44(0) 1892 53 0358

ISBN : 978 1 906574 06 2

British Library Cataloguing in Publication Data

A Catalogue record for this book is available from the British Library

Every effort has been made to make the book error free. However, the author and publisher have no warranty of any kind, expressed or implied, with regard to the documentation contained in this book.

Preface

The present technological scenario is fast changing. New materials and products are being developed very rapidly. These developments are taking place to meet the ever increasing demand for better quality products, greater efficiency, increased service life and stringent reliability requirements. In view of these requirements of today's technological society, the importance of the knowledge of Non-Destructive Testing (NDT) techniques has increased considerably.

As the name suggests, Non-Destructive Testing techniques include all types of testing techniques in which the future usefulness of the material under test is not impaired. In simpler words, the material under test is not destroyed and may be reused. This is not so in "Destructive Testing" such as tensile testing, impact testing, burst pressure testing etc., wherein the sample or the product under test is destroyed and cannot be reused.

Sometimes Non-Destructive Evaluation (NDE) is used instead of NDT and many people use 'NDE' and 'NDT' interchangeably. However, it would be more appropriate to call only those techniques as Non-Destructive Evaluation (NDE) techniques which are primarily meant for evaluation of material properties such as fibre volume fraction in fibre composites, composition of alloys, grain size, tensile strength, fatigue behaviour etc. nondestructively. Other techniques which are conventionally used for inspection of incoming materials/stores or for inspection of structures, for location of manufacturing or in-service defects (cracks, voids, porosities etc.) should be called NDT techniques.

Use of NDT techniques results in better understanding of the material behaviour and it allows the separation of defect-free and defective materials/products. Hence, use of NDT techniques results in increased confidence level of designers in the material to be used and this in turn allows the designer to select an appropriate lower 'factor-of safety' value and thus providing a far more compact and reliable design with added advantage of material saving.

The needs and benefits of NDT techniques, nature of flaws which are detected by NDT techniques, various applications of NDT techniques, types of NDT techniques, different steps involved in any typical NDT technique and use of NDT techniques for applications other than flaw detection have all been described in the very first chapter of the book.

The main aim of this book is to provide the much needed basic concepts of various NDT techniques in simple language. Once a user understand the concepts of these NDT techniques, he will be in a better position to use them properly and interpret results correctly. This will in turn lead to making NDT techniques more popular.

Besides serving the primary purpose of providing a textbook in the field of Non-Destructive Testing, it would also provide the much needed reference book to various Engineers and Research Scientists who use NDT techniques for inspection purposes or for material behaviour study or for their research.

The following NDT techniques have been described in detail in this book:

1. X-Radiography
2. Ultrasonic Testing
3. Acoustic Emission Testing and Acousto-Ultrasonic Testing
4. Eddy-Current Testing
5. Magnetic Particle Flaw Detection
6. Liquid Penetrant Inspection
7. Miscellaneous methods such as Optical Methods, Pressure and Leak Testing, Sulphur Printing, Spark Testing, Thermal Methods, Spectrochemical Methods, Electrical Methods, Beta Gauge Technique etc.

The author wishes to most sincerely acknowledge all those who have directly or indirectly helped the author in the publication of this book. The author would welcome any constructive suggestions from the valued users of this book and especially so from his fellow colleagues in the NDT field.

RAVI PRAKASH

Contents

CHAPTER **1**

Introduction

In today's rapidly growing industrial world where the requirement of reliability is increasing day by day and where newer and advanced modern materials are being introduced on a large scale, Non-Destructive Testing (NDT) techniques have a very important role to play. NDT techniques are used for the evaluation of defects in various materials/components and for the characterisation of material properties. Use of NDT techniques leads to better understanding of material behaviour and this in turn leads to increased confidence in the material being used. Higher confidence means lower factor of ignorance and the designer can thus afford to opt for a lower value of the factor of safety without sacrificing the reliability. Lower value of factor of safety results in decreased dimensions, leading to saving in weight and the material as such. NDT techniques are also used for routine/periodic inspection of various industrial processes/structures and lately new NDT techniques have been developed for continuous monitoring of critical industrial structures/processes. NDT techniques are practically used in every engineering industry and widely used in more demanding industries such as aircraft industry, space industry, nuclear establishments, power plants and chemical/fertiliser plants etc.

1.1 DEFINITION OF NON-DESTRUCTIVE TESTING

Non-Destructive testing may be defined as those testing methods in which the material under test is not destroyed or to say that the future usefulness of the material under test is not impaired. To explain it further, one may say that unlike mechanical testings (*e.g.,* tensile testing, flexural testing, torsional testing etc.) in which the material under test is made to fail (fractured into two or more pieces) to evaluate the strength whereas in non-destructive testing, the material under test is not destroyed at all and the material under test retains all its original properties (*i.e.,* after the test, the material or component under test can be used for the purpose it was originally intended). To drive the point further home, suppose one wants to evaluate the burst strength of a newly fabricated pressure vessel and if a destructive testing method such as hydrostatic failure test is opted for, one would be able to evaluate the burst strength but only after loosing the pressure vessel. On the contrary if a non-destructive technique is chosen for inspecting the flaws in the pressure vessel to make sure that it would withstand the calculated pressure or the designed pressure, one may be assured of the burst strength without causing any harm to the pressure vessel, what-so-ever. New NDT techniques can evaluate the burst strength directly too without searching for defects as such.

1

Sometimes NDT techniques are also used for the characterisation of physical/mechanical properties of various materials. For example, one may measure the Young's modulus of elasticity by measuring the ultrasonic velocity in materials; one may evaluate the ultimate tensile strength, bond strength of an adhesively bonded joint or burst pressure of a pipe/pressure vessel by measuring the stress wave factor; one may evaluate fibre volume fraction in carbon fibre reinforced plastic composites by measuring the changes in the impedance of an eddy-current probe; one may evaluate the grain size using ultrasonic attenuation measurement etc. When NDT techniques are used for such applications, they are referred to as Non-Destructive Evaluation (NDE) techniques. In fact NDE techniques cover all the mechanical characterisation techniques as well as the NDT techniques which are used for defect location and evaluation. Hence, NDE is a far wider term as compared to NDT. However, some people use the terms NDE and NDT interchangeably and it has now become an accepted practice to use either NDT or NDE to describe all the testing techniques in which the future usefulness of the material or product is not impaired.

Besides these abbreviations, *i.e.,* NDT and NDE, certain authors and inspection engineers use yet another abbreviation, *viz.,* NDI (Non-Destructive Inspection) to describe all NDT techniques which are used to inspect the incoming raw materials/products and the techniques which are used for routine/periodic inspection.

1.2 NEED FOR NDT TECHNIQUES AND ITS APPLICATIONS

Materials Scientists and Materials Engineers know that no material can be categorised as absolutely perfect (*i.e.,* having zero defect). One can only minimise the amount of defect or may reduce the defect size by proper selection of manufacturing processes or by improving the production processes. As one has to live with the defects, he/she should be able to locate and assess the severity of the defects present in any material, component, product, system or plant without impairing their future usefulness. To meet this requirement, various NDT techniques are used.

Nowadays when everyone is going for higher reliability and stringent quality checks, the NDT techniques have gained all the more importance. Whether one talks about quality assurance or quality control or reliability, one has to understand the science and engineering of various NDT techniques to form a firm base. Also, in view of the newer materials being developed at a very rapid rate to meet the diverse requirements of modern engineering industries (such as very high specific stiffness, high conductivity, low attenuation etc.), NDT techniques have a far greater role to play. Most of the advanced composites and ceramic materials and to a lesser extent, polymeric materials too have inherent defects and their non-destructive testing becomes a must, to develop confidence in these newer advanced materials. Using NDT techniques one may select relatively "defect-free" materials and better understand the behaviour of newly developed materials. This better understanding reduces the so-called "factor of ignorance" and thereby allows the designer to go for a lower "factor of safety" without sacrificing reliability of the product. Lower values of factor of safety result in reduced cross-sectional area, lighter components and material saving. Thus, there exists a strong relationship between NDT and the design process.

NDT techniques are used at various stages in an engineering industry. Generally all incoming materials/components are inspected nondestructively before their acceptance by the stores. Components are also inspected during and after manufacturing and often during their service life to ensure that the condition is suitable for the purpose. A raw material or component, which has been checked nondestructively before supplying the same, merits a premium, when assured of its quality and reliability. It would be a better practice specially, if at the inquiry/ quotation/tender level itself, NDT requirements are clearly specified (*i.e.*, non-acceptable defects are clearly mentioned), alongwith other technical specifications.

NDT techniques assume greater significance in high reliability sectors such as nuclear, space, aircraft, defence, automobile, chemical and fertilizer industries. NDT techniques are also widely used for power equipments, heat exchangers, pressure vessels, heavy engineering items etc. In fact, practically all large and medium scale engineering industry uses NDT techniques in one form or the other.

Beside being used for defect evaluation and location, NDT techniques are used for assessing the severity or otherwise of the defects too. Whereas certain defects may be totally harmless in view of their small size or innocuous position (*e.g.*, microvoids in a low-stressed region of a composite), others may be quite harmful in view of their large size, type and position. These harmful defects grow to dangerous proportions in service but the harmless defects, as their name suggests, may be ignored. In addition to *harmful and harmless* categories, there is a third category of defects too. They are classified as *beneficial defects*. These defects are not severe and they are beneficially located. Certain beneficial defects arrest or impede the progress of a propagating crack and certain other beneficial defects divert the path of crack progress by providing an easier path but in a different direction which is relatively safer. Hence, the NDT engineer should not reject a material/component merely because of the presence of a defect. The NDT engineer should find out whether the defect is harmless or beneficial and categorically state the rejection criteria.

With the advent of modern advances in the field of electronics, computers and data processing; greater use of electronic gadgets, computer interfaces and data processors are being made to improve the existing NDT techniques. This has enabled development of instruments and softwares which quickly detect the defects and identify their nature, shape, size and criticality. Another current trend in the field of non-destructive inspection is to continuously monitor high risk structures subjected to service loads (*e.g.*, acoustic emission monitoring of pressure vessels and storage tanks). This continuous monitoring results in higher reliability of structures/components in service and permit prolonged service life of the structures/ components. In addition to this continuous in-service monitoring, NDT techniques are also being used for on-line monitoring during manufacturing too. (*e.g.*, on-line monitoring of welding defects during welding operation using acoustic emission technique).

As mentioned in the previous section, NDT techniques are used for the evaluation of material behaviour too. One may study in detail various stages leading to ductile failure of different metals or different stages leading to failure of fibre reinforced plastic composites etc. using various NDT techniques. One may use NDT techniques for predicting fatigue behaviour of different materials, for evaluation of different mechanical properties (modulus of elasticity, tensile strength, burst-strength etc.), for evaluation of in-service embrittlement, for evaluation of grain size in various alloys, to quantify composition of alloys, to differentiate between different metals and alloys (*i.e.*, identification of metals and alloys), for measurement of fibre volume fraction in fibre reinforced plastic composites etc.

Yet another emerging usage of NDT techniques is in the field of engineering "postmortem", *i.e.*, for analyzing the test data after a simulated failure or after the proof loading. Fractured surfaces are also studied for probing the reasons for failure. Material composition at the fracture site, texture of fractured surfaces, fibre pull-outs, presence of defects (cracks, voids, inclusions etc.) on the fracture surfaces, all help in proper investigation of the cause of failure. This engineering postmortem is done with the help of various microscopic techniques (including scanning electron microscopy) and NDT techniques. The results of such investigations help the designers and/or production engineers to improve the design and/or production/fabrication processes involved.

Application of NDT techniques are many and new applications are being reported frequently. The field of non-destructive testing has been growing steadily in the past and has resulted in increased safety and cost savings. The future of NDT techniques is even brighter especially for the characterisation of newer materials such as smart materials, nano-materials and advanced ceramics.

1.3 TYPES OF NDT TECHNIQUES

There are a number of NDT techniques which exist today in various industries/organizations. Some of these NDT techniques are for certain specific applications to suit the requirement of a particular industry, whereas other NDT techniques are more broad based and may be used for varied applications. The common NDT techniques are as listed below:

1. Liquid penetrant inspection

2. Magnetic particle flaw detection

3. Radiography (X-ray, Gamma ray and Neutron)

4. Ultrasonic testing

5. Eddy-current testing

6. Acoustic emission and acousto-ultrasonic testing

These NDT techniques and certain other miscellaneous NDT techniques have been described in the Chapters to follow. Besides these general purpose or conventional NDT techniques, there are certain other NDT techniques used by industry for certain specific applications. These are Corona discharge, microwave, dielectric methods, microradiography, chemical and Mössbauer spectroscopy, vibration techniques, acoustography, eddy-sonic technique, computer tomography, X-ray diffraction, neutron diffraction, magnetic rubber, optical and laser techniques; gas, air and helium leak detection techniques; radiography with high density fillers, holographic interferometry, etc.

1.4 BENEFITS FROM NON-DESTRUCTIVE TESTING

The advantages of using NDT techniques are many. These benefits have been and are being more fully recognized by the engineering industries as a means of meeting the consumer demands for better quality products, reduced cost and increased production. Quality of products improve because by using suitable NDT techniques, one is able to identify unacceptable material for production in the very beginning and one is able to inspect the product for manufacturing defects etc. after completion or during the production process itself. Hence, use of non-destructive testing techniques, completely eliminates the chances of a faulty or sub-standard product reaching the consumer. This in turn leads to customer satisfaction and higher reliability of the product.

Use of NDT techniques results in increased productivity and higher profit. Inspection at different stages prevents wastage of material by way of less scrap and it also minimises the loss of manpower by way of less rework. By following the above factors, higher productivity and higher profits are achieved. Defect-free components lead to highly reduced service cost too.

NDT techniques help design engineers in locating regions of high mechanical stresses and potential sites for fatigue cracks etc. This provides necessary guidelines to design engineers for improving the design accordingly. NDT techniques help the production engineers too because by monitoring various production processes using suitable NDT techniques, they can pinpoint trouble spots and accordingly bring in desired improvements. NDT engineers can also ascertain the type of defect, its size, its orientation etc. and decide whether or not the defect is harmful. If it is not harmful, the material/product need not be rejected. Moreover, by locating a defect, only that portion of the material needs to be removed. Hence, by using NDT techniques some defective materials may become usable and thus salvage of material becomes possible.

By monitoring production processes and by regular inspection, one may prevent malfunctioning of key equipments/processes and probably one may completely eliminate the breakdown of various equipments/processes.

Efforts are currently being made by various industrial houses and research organisations to develop on-line monitoring NDT techniques *i.e.,* NDT techniques which can inspect the components/products etc. during manufacturing/service. On-line monitoring of machines, engines, turbines, railways etc. would avoid shut-downs for routine inspection and thus overhauls would become less frequent. In short, on-line monitoring would result in very significant savings in cost and time, increased productivity and zero risk of catastrophic failures. Latest NDT techniques such as Acoustic Emission Testing and Acousto-Ultrasonic Testing hold great promise for development of suitable on-line continuous monitoring for various structures, components, systems etc.

As the use of NDT techniques leads to much higher safety standards (*i.e.,* prevention of accidents, prevention of loss of life and prevention of loss of property), their importance cannot be denied by any logic. Use of NDT techniques for regular inspection and continuous monitoring of various processes/structures is required by various safety ordinances. Certain old ordinances/ by-laws need updating to introduce new NDT techniques for better safety of various industrial processes/structures.

NDT techniques also help in material sorting, for evaluation of chemical composition, for measuring differences in physical and metallurgical properties, for ascertaining proper heat treatment and for studying the mechanical behaviour of various materials. This helps different engineering industries in different ways. NDT techniques also help in the determination of residual strength and remainder life.

1.5 NATURE OF FLAWS

NDT techniques in general detect any inhomogeneity present in the material under inspection. If these inhomogeneities are harmful they are categorised as flaws or defects. There are various types of flaws encountered by the NDT engineers during inspection of incoming raw materials and/or finished products. These flaws can be broadly categorised into three different groups *viz.* inherent flaws, processing flaws and service flaws. The first category of flaws, *viz.,* inherent flaws are those which are present in the incoming raw material. These flaws are introduced during the initial production of raw material. Some of inherent flaws are porosities, blow-holes, voids, delaminations, seams, flakes, shrinkage or thermal cracks, segregation, rolling and plating defects, inclusions, surface-cracks etc. In the field of fibre reinforced plastic composites, inherent flaws are voids, shrinkage cracks due to thermal anisotropy of different 'pre-preg' layers (*e.g.,* in case of carbon fibre reinforced plastics), fibre-wash or fibre-kinks, resin rich areas, delaminations, inclusions such as silicon particles from silicon mould release agent, etc.

The second category of flaws *viz.,* processing flaws occur during the processing, manufacture or assembly of components. For example, if welding technique is used for either manufacture or for assembly and one is not very careful about its correct usage, it may introduce flaws such as slag, gas-holes, heat-affected zones, cracks due to bad geometry, residual stresses and distortion etc. Similarly, if it is an adhesively bonded assembly, lack of bond may be introduced leading to imperfection of the final product. Machining marks or grinding defects may be introduced during machining and grinding of components. These machining defects are generally surface defects and lead to fatigue failure of the component during the service life of the component. Improper machining may lead to corners which act as stress-concentrators and which are potential crack nucleation sites. Heat treatment processes also sometime give rise to thermal defects, embrittlement, decarburization etc.

The third category of flaws, *i.e.,* service flaws are introduced during the service life of the material or component or structure. Fatigue, corrosion, stress corrosion, wear, embrittlement, corrosion fatigue etc. are examples of service flaws. Out of these flaws, corrosion is the most obvious but majority of structural failures during service are caused by fatigue.

Nature of flaws described above is on the basis of stage at which they are introduced. However, if one classifies the flaws, depending upon the position, they can be categorised into three types *viz.,* surface flaws, sub-surface flaws and internal flaws. As their names suggest, surface flaws are open to surface and internal flaws are present within the volume of material. Sub-surface flaws are very near to surface of material/component but they are not open to surface. Surface cracks, pits, tool marks, grinding cracks, fatigue cracks etc. are all surface flaws whereas blow-holes, porosities, voids, inclusions, delaminations etc. are internal cracks. Fatigue cracks generally originate from the surface and have therefore been listed along with other surface cracks. However, if gross-defects are present within the volume of the material, fatigue cracks may initiate at these internal features also and may propagate within the bulk of the material. In such cases, it becomes internal flaw. In fibre reinforced plastic composites too, they mostly originate from surface or from surface flaws. However, in case of gross defects (voids, delaminations, resin rich areas or thermal cracks) being present, fatigue cracks originate from these defects and are treated as internal flaws.

1.6 VARIOUS STEPS INVOLVED IN NON-DESTRUCTIVE TESTING

The following steps are usually involved in any non-destructive testing schedule:

1. Preparation of test surface
2. Application of testing medium/signal
3. Modification of testing medium/signal
4. Conversion of the modulated or changed medium/signal into a convenient form
5. Interpretation of results obtained
6. Verification of test results.

In first step, any material/product under test is properly cleaned for inspection in most cases. However, in certain cases, one may be required to clean only a small portion of test surface to enable a NDT probe/transducer to be placed over it. If test surface is not cleaned properly, it may give rise to spurious indications causing confusion. In fact, it would be a great help to NDT engineers if design engineers make a provision for a good surface finish in a local area at the design stage itself and thus making the proper inspection possible and would increase its effectiveness. In the Chapters to follow, various cleaning methods have been described. One may use wire brush, cloth and cleaning fluid for smaller components and may be required to go for shot blasting etc. for bigger components. Generally speaking, rust, dust, grease, paint etc. should not be present on the surface of the object to be inspected.

The second step involves application of testing medium or testing signal. For example, in radiography, X-rays are the testing medium; in visual testing, visible light is the testing medium; in magnetic particle inspection technique, magnetic field is the testing medium; and in acousto-ultrasonic technique, incident ultrasonic signal is the testing signal. So depending upon the technique being selected, testing medium changes and one can go on citing several examples of testing medium.

Once the testing medium/signal is applied, specimen under test modifies the testing medium or one may say that the applied signal gets modulated depending upon the quality of specimen. This step is the third step involved in any NDT test. For example, in radiography, any defect in the material under test modifies intensity of radiation reaching the X-ray film on opposite side of test material; in optical method, intensity of transmitted light is modified by any flaw or by any marking on a transparent sample; in magnetic particle inspection, magnetic field gets distorted at flaws and produces a leakage flux; and in acousto-ultrasonic testing, the ultrasonic pulse is modulated or dampened by the presence of any flaw. These examples illustrate the step: "modification of testing medium/signal".

The next step involves conversion of modulated signal or changed medium into a convenient form to facilitate the interpretation of test results. For this step, a suitable detector is employed, which detects changes or modifications in test medium/signal. For example, in X-radiography, X-ray film acts as a detector; in optical method, eye or a photometer acts as detector, in magnetic particle inspection technique, the magnetic particles act as detector and form suitable pattern for easy detection, and in acousto-ultrasonic technique, an acoustic emission sensor is used as detector for modulated ultrasonic signal.

The fifth and mostly the last step involved in any non-destructive testing schedule is interpretation of test results. This is most important step and this is the step where skill and experience is mostly needed. In NDT tests, it is relatively easier to obtain test signal but it is not so easy to correctly interpret the test results. For example, in X-radiography, by looking at variations in darkness on the radiograph, NDT engineer has to tell, whether or not, flaws are present and what type of flaws are present. In optical method, if visual testing is involved, the opthomological nervous system converts the intensity variations observed by the eye into suitable impulses and the brain subsequently does the proper interpretation. If a photometer is used instead, the digital read-out of a DVM connected to the photometer provides necessary information and a trained NDT engineer properly interprets it. Similarly, in magnetic particle inspection technique, the magnetic particles adhere to the test object. Proper interpretation involves whether it is due to presence of a crack (surface or sub-surface) or is it simply because there is a change in the cross-section or could be that the particles are adhering to edges only. Wrong interpretation may lead to rejection of sound material resulting in material wastage and loss. Thus, a NDT engineer who interprets the results determines the success or failure of a NDT test.

As far as verification of test results are concerned, it is done with the help of some destructive methods and sometimes needed for convincing the supplier or the fabricator to show that the test results have been correctly interpreted. Verification of test results is also required when new NDT techniques are developed. Even during routine and conventional NDT tests, periodic destructive tests should be applied to make sure that the tests and associated instruments are operating properly. The resolution of any technique and its sensitivity is also thus gets checked.

When new NDT techniques are developed, destructive tests are used for calibration purposes too. For example, as has been described in Chapters to follow, burn-off technique or quantitative microscopic technique can be used to destructively evaluate the fibre volume fraction in carbon fibre reinforced plastic composites and when correlated with the meter reading of the eddy-current system, one gets the calibration curve. Similarly, one has to go for destructive micrographic study to evaluate defect percentage in FRP samples to correlate the same with ultrasonic attenuation of fibre composites. There are many other examples for which, destructive tests are used for evaluating certain parameters to obtain calibration curves.

1.7 USES OF NDT TECHNIQUES FOR APPLICATIONS OTHER THAN FLAW DETECTION

As was mentioned earlier in this Chapter, NDT techniques are used not only for flaw detection but are also used for number of varied applications. The most common usage of NDT technique for applications other than flaw detection is in the field of thickness measurement. Pulse-echo method of ultrasonic testing can easily be used for accurate measurement of thickness. Commercially available ultrasonic thickness gauges provide quick and accurate thickness readings. Thickness gauging can also be done by adopting radiation absorption measurements. Beta-rays and Gamma-rays can be used as radiation sources and based on this Beta-ray and Gamma-ray; thickness gauges have been developed. Besides these, alpha and beta backscatter gauges also exist for thickness measurement. Thickness gauge based on eddy-current principle also exists.

Besides thickness measurement, NDT techniques are also used for applications such as identification/classification of materials, evaluation of chemical composition, evaluating the surface characteristics and surface finish, locating the areas of stress concentration and critically loaded areas, for evaluating physical and mechanical properties, for evaluating bond strength, for checking in-service embrittlement, for studying material and fracture behaviour, for use as an incipient failure detection system (IFDS) and for measuring residual stress etc.

For identification/classification of materials, one may site following NDT techniques: spark test, spot test, sulphur printing, eddy-current testing, ultrasonic testing etc. For cross-plied carbon fibre reinforced plastic composites, eddy-current method can be employed for identification of lay-up order.

For evaluation of chemical composition, one may go for spectrochemical analysis and for surface characteristics study, one may use nuclear scattering techniques. Using techniques of photoelasticity and Moire method, one can find the areas of stress concentration and critically loaded areas. Brittle lacquer can also be used for such applications. In fact, by selecting a brittle lacquer of proper threshold strain value, one can get information about the impending failure because if threshold strain of the lacquer is lower than failure strain of the component/structure, lacquer will fail earlier than the component/structure and would thus provide a warning about the impending failure. Even acoustic emission system provides proper warning about the impending failure and the technique can be used as an incipient failure detection system.

For evaluating physical and mechanical properties, many NDT techniques are used. For example, ultrasonic technique can be used for evaluating grain size, modulus of elasticity, optimum dwell time for reinforced plastic mouldings, residual stresses, etc. Eddy-current testing technique can be used for evaluation of fibre volume fraction in carbon fibre reinforced plastic composites, acousto-ultrasonic techniques can be used for evaluation of ultimate tensile strength, fatigue strength, burst pressure of pipes and vessels, bond strength of adhesively bonded structures etc.

Acoustic emission technique can be used for checking in-service embrittlement and for studying deformation of materials and fracture behaviour of different materials. Details of these applications have been provided in the concerned chapters.

1.8 CONCLUDING REMARKS

From the preceding sections it is quite obvious that the main purpose of non-destructive testing is to determine the suitability of a particular part or a system, to satisfactorily perform its intended function. With this purpose, the field of non-destructive testing can be a significant contributor to increased efficiency and utilization of machines and materials as well as a key item in assuring safe performance. Structural failures in big passenger airliners, offshore oil platforms, nuclear reactors, chemical and fertilizer plants etc. remind us of the cost in resources and human life resulting from material fracture. Effective NDT coupled with fracture mechanics analysis, can be used to avoid such catastrophic failures. The present emphasis on quality, reliability and life extension of plants and machinery requires full utilization of various NDT techniques.

Besides being used for high technology products such as nuclear reactors, supersonic aircrafts, spacecrafts, rockets and missiles etc., NDT techniques are finding application in the field of inspection of more consumer-oriented products and in the field of Biomedical Engineering too, *e.g.*, inspection of pacemakers, parts of surgical implants, bone plates, external fixators, spinal rectangles etc.

CHAPTER 2

Ultrasonic Testing

Ultrasonic testing is one of the most popular non-destructive testing techniques for the detection of internal flaws. Using ultrasonic testing technique, besides evaluating internal flaws, one could detect surface and sub-surface discontinuities, measure thickness where ordinary means of thickness measurement fail, evaluate physical properties such as elastic modulus, study metallurgical structure such as grain size etc. Because of its high penetrating power, one could inspect extremely thick sections and because of its high operating frequency, it is possible to detect very minute cracks too. There are two basic types of ultrasonic testing, *viz.,* pulse-echo technique and through transmission technique. Both the techniques have their own advantages and limitations. For pulse-echo technique, there is need for access to only one surface of the specimen; whereas for through transmission technique, access to both sides of the specimen is necessary. Ultrasonic technique is very versatile indeed and besides being widely used for various industrial applications this technique is also being extensively used for various research applications. Also ultrasonic testing does not cause any health hazard to the operator or to the persons standing nearby.

2.1 INTRODUCTION

Ultrasonic testing techniques are widely used in industry for inspection of incoming materials as well as for quality control. The flaws to be detected may be voids, cracks, inclusions, segregations, delaminations, bursts, flakes or similar type flaws. These flaws may either be inherent to the raw material or may result from fabrication and heat treatment. The flaws may also occur during the service time from fatigue, creep, corrosion and similar causes.

The principle of ultrasonic testing is based upon the interaction of sound waves with the internal structure of the material. The sound waves pass through a defect-free material relatively easily as compared to a material having defect because at defects, sound waves get reflected and/or scattered resulting in loss of energy.

To explain the principle of ultrasonic testing, there is an oft quoted example of cracked bell. When an uncracked bell is struck, it sounds very differently as compared to the sound, which shall emanate from a cracked bell. The ringing sound of uncracked bell, dampens very rapidly. Hence, by listening to the sound from the bell, one can say whether or not the bell is cracked. In this example, the crack in question is a gross crack and detection of such cracks is possible by sound waves in the audible frequency range. For detection of minute cracks, however, one has to go for sound waves having frequencies higher than that of audible sound waves.

For "higher" there is a Greek word "ultra" and for sound there is a Greek word "sonic" and, therefore, the waves which have frequencies higher than the frequency of audible sound waves are called "ultrasonic" waves.

2.2 FREQUENCY OF ULTRASONIC WAVES

As already described, ultrasonic waves have frequencies higher than the frequency of audible sound, *i.e.,* frequencies higher than 20 kilocycle per second (20 kHz). However, most commercial ultrasonic testing is done at frequencies in Megacycle per second (MHz) range, *viz.,* 1 to 25 MHz. However, there also exist applications requiring frequencies as low as 25 kHz and as high as 200 MHz.

Generally, the choice of test frequency depends upon two factors, *viz.,* the minimum size of defect, which is to be detected and the medium in which such a defect is situated. As an example, let the minimum size of defect, which is to be detected be of the order of 0.5 mm and the velocity of sound in the medium in which the defect is situated, be of the order of 5000 m/sec. Also, as is well known, for a defect to be easily detected, the wavelength of ultrasonic waves should of the order of the size of defect. Therefore, using the well known relationship between velocity, frequency and wavelength, *viz., $v = f \cdot \lambda$* (where v = velocity, f = frequency and λ = wavelength)

$$f = v/\lambda$$

$$= \frac{5000 \times 10^3 \text{ mm sec}^{-1}}{0.5 \text{ mm}}$$

$$= 10 \times 10^6 \text{ cycle/sec. or } 10 \text{ MHz}$$

In this example, if one would like to detect cracks of the order of 1 mm only and would not bother for finer cracks, a test frequency of 5 MHz (*i.e.,* a lower test frequency) would suffice. However, for detecting finer cracks (say of the order of 0.25 mm), higher test frequency (20 MHz) would be required. This example also demonstrates as to why, generally a test frequency in MHz range instead of kHz range is selected for detecting particle size defects in engineering materials.

2.3 GENERATION OF ULTRASONIC WAVES

As is widely known, all mechanical vibrations, either for analysis or for test purposes, are generated by electromechanical transducers. The electromechanical transducers transform electrical energy into mechanical energy and vice versa. For generation of ultrasonic waves, *i.e.,* for frequencies of mechanical vibration above 200 kHz, piezo-electric transducers are used. For ultrasonic non-destructive testing, though the frequency of vibration is high, the amplitude of vibration is low. Because of this low amplitude vibration, ultrasonic tests do not affect the component under test. This is unlike high amplitude forced vibration destructive tests, which may result in heating, permanent deformation or failure of the component under test.

Piezo-electric transducers employ materials which generate electric charges when mechanically stressed. Conversely, piezo-electric materials become stressed when electric charge is applied to them. Piezo-electric materials have been briefly described in the next section. These piezo-electric materials are suitably mounted for ease of using them for ultrasonic inspection. The mounted units are called ultrasonic probes. These ultrasonic probes generate a short train of ultrasonic waves when an electrical pulse is applied to piezo-electric crystal housed inside the probe. This short wave train is directed into the material under test by means of proper couplant. This couplant can be solid or liquid. Couplant is also used to get a particular type of wave. The received ultrasonic wave (reflected or transmitted) also consists of a series of short wave train. This wave train is detected by the receiver transducer (which could be same as transmitting transducer), which produces an electrical analogue signal of this wave train. Hence, the processed signal, which is displayed on the screen of ultrasonic flaw detector is not the wave train but an envelope of the wave train.

2.4 PIEZO-ELECTRIC MATERIALS FOR ULTRASONIC TRANSDUCERS

There are certain natural and certain man-made piezo-electric materials which expand and contract under the influence of varying electric fields. These piezo-electric materials provide an electrical output signal on being strained mechanically.

There are a number of piezo-electric materials which are currently being used for ultrasonic transducers or probes. Out of these, quartz, barium titanate and lead zirconate titanate (PZT) are most common. Quartz is a natural piezo-electric material whereas the rest two are so-called "man-made" piezo-electric materials. A brief description of transducers made from these piezo-electric materials are as follows:

(*i*) **Quartz transducers:** Quartz is the most widely used piezo-electric material. The reasons for its wide use are its high stability, insolubility, chemical inertness, resistance to ageing and ability to withstand high temperatures. However, quartz crystals are incapable of generating high acoustic power and they suffer from mixed mode deformations.

(*ii*) **Barium titanate transducers:** Barium titanate is a man-made ceramic piezo-electric material and is one of the most effective ultrasonic generating material. It is, however, a poor receiver of ultrasonic waves, *i.e.*, if strained, it produces a much lower voltage as compared to a voltage which would have been produced by applying equal strain to, say, quartz of PZT crystals.

(*iii*) **Lead zirconate titanate (PZT) transducers:** Lead zirconate titanate, a ceramic material, is widely used for making ultrasonic transducers. The transducers made out of this crystal are marked PZT on top. This is a polycrystalline material and when it is polarised by applying a strong electric field, it starts behaving like a piezo-electric material. Using this piezo-electric material, transducers of many shapes can be produced. These transducers have little effect of moisture on them and can withstand temperatures up to about 100°C. The impedance of lead zirconate titanate crystals is low as compared to natural piezo-electric crystals like quartz and, therefore, the voltage source required for exciting PZT transducers is lower than that required for quartz transducers.

2.5 DIFFERENT KINDS OF ULTRASONIC TRANSDUCERS

(*i*) **Straight beam contact ultrasonic transducers:** Straight beam contact ultrasonic transducers have crystals with electrode on one face and the other face remains exposed (*i.e.*, the part under test acts as the second electrode). Electrodes are there to provide a voltage gradient across the crystal element. Due to this voltage gradient, acoustic energy is developed in the crystal. The voltage gradient is generally very high and is of the order of about 5000 volts per mm and that is why, adequate insulation is always provided. This voltage gradient usually takes the form of a large (100 to 2000 volts) pulse lasting for less than 10 m sec. Gold and silver electrodes are usually vapour deposited on the desired side of the piezo-electric crystals. In certain applications, instead of vapour depositing, thin metallic foils are cemented to the piezo-electric crystals.

The straight beam contact ultrasonic transducers can be used for testing any reasonably flat surface which is a good conductor of electricity. Such transducers are widely used in the frequency range of 500 kHz to 10 MHz.

(ii) **Straight beam faced ultrasonic transducers:** This type of ultrasonic transducers can be used for testing even rough surfaces and which may either be good conductor or bad conductor of electricity. In such transducers, electrodes are provided on both the sides of piezo-electric crystal. In order to protect the front electrode from breaking and thereby avoiding the breaking of piezo-electric crystals, wear plates are provided. These wear plates are generally made of perspex (acrylic). However, other plastics may also be used as wear plates. The wear plates are also sometimes referred to as face plates. In certain applications, instead of plastic wear plates, metal or rubber wear plates are also used. Curved wear plates are used for inspection of tubes and pipes.

(iii) **Angle beam contact ultrasonic transducers:** Using a plastic wedge between the piezo-electric crystal and the part under test, it is possible to direct the ultrasonic beam towards a particular area in a part and not in the direction normal to the surface (Fig. 2.1). Generally, perspex is used as the wedge material and these transducers are widely used in the frequency range of 1 to 5 kHz. By suitably selecting the wedge dimensions, one may either get shear waves due to refraction or surface waves due to mode conversion. One may also use curved wedges to suit pipe inspection.

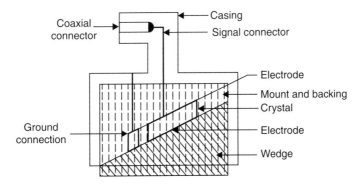

Fig. 2.1 Angle beam contact ultrasonic transducer

(iv) **Immersion type ultrasonic transducers:** As shall be described later in section 2.12, there are two types of ultrasonic test methods commonly employed for non-destructive testing of various materials and workpieces. These are pulse-echo and through transmission methods. Whereas in pulse-echo method, a thin layer of grease is ordinarily used as couplant; in through transmission testing, transducers are usually a fair distance apart from the workpiece and the transducers are acoustically coupled to the workpiece through water. For such applications, one requires immersion type ultrasonic transducers. In these transducers, piezo-electric crystal's mount and backing materials and all the electric connections are properly waterproofed. Also, as shown in Fig. 2.2, a grounding electrode is provided on the front face. Such transducers are available in the frequency range of 0.5 MHz to 20 MHz.

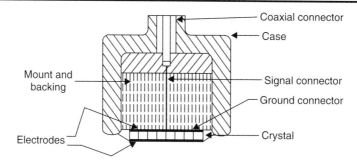

Fig. 2.2 Immersion type ultrasonic transducer

(*v*) **Focused ultrasonic transducers:** Sometimes, immersion type ultrasonic transducers have an acoustic lens attached to the front face of the transducer, making it a focused transducer. If the lens has spherical curvature, it provides point focus. The focusing of beam provides a higher beam intensity. Focused ultrasonic transducers can be compared with optical magnifying lens. Just like a magnifying lens views a small area and provides the details of the area; focused transducers also inspect a small area because of their higher beam intensity, provide the information about that area in greater details. Commercially available focused ultrasonic transducers generally provide a long and narrow (cigar shaped) ultrasonic beam. In such transducers, most desirable portion of the ultrasonic beam starts at the point of maximum intensity and extends for a considerable distance beyond this focus point. Such transducers are used for inspection of thin materials and for the evaluation of bond quality of laminar sandwich type materials.

(*vi*) **Longitudinal wave ultrasonic transducers:** These ultrasonic transducers produce ultrasonic beams in such a way that the particles of medium under inspection, vibrate along the direction of beam travel (Fig. 2.3). Longitudinal wave transducers are also known as compressional wave transducers. Straight beam contact ultrasonic transducers, straight beam faced ultrasonic transducers and immersion type ultrasonic transducers described earlier are also longitudinal wave transducers, if classified according to the wave type. Frequency of these transducers is either inscribed over the transducers casing or internationally accepted colour coding is used.

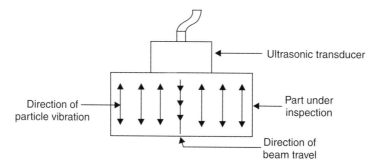

Fig. 2.3 Longitudinal or compressional wave ultrasonic transducer

(*vii*) **Transverse wave ultrasonic transducers:** Transverse wave ultrasonic transducers produce ultrasonic beams in such a way that the particles of medium under inspection vibrate perpendicular to direction of beam travel (Fig. 2.4). These transducers are popularly known as **shear wave transducers**. Angle beam contact ultrasonic transducers described earlier are nothing but transverse wave or shear wave transducers when defined with wave propagation characteristics in mind. These transducers are generally available in the frequency range of 1 to 5 MHz and the usual direction of beam travel is at an angle of either 30° or 45° or 60° with respect to the perpendicular to transducer's face. These angles are usually inscribed on the transducer's casings.

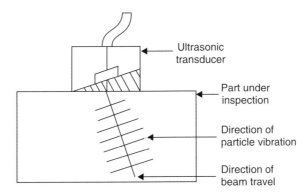

Fig. 2.4 Shear wave or transverse wave ultrasonic transducer

(*viii*) **Surface wave ultrasonic transducers:** Besides longitudinal and transverse waves, surface wave is the third principal mode in which ultrasonic beam travels. In these surface wave ultrasonic transducers, the angle of incidence is so selected that it is greater than or equal to the critical angle. Critical angle can be determined using well known equation for determination of critical angle and for angles equal to or greater than the critical angle, no reflected or transmitted beam exist. Only beam at the surface is found due to total reflection. These transducers have either "SF" inscribed over them or sometimes 90° is inscribed to indicate that there is an angle of 90° between the direction of beam travel and the perpendicular to the transducer's face.

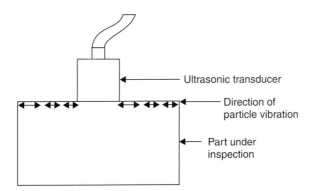

Fig. 2.5 Surface wave ultrasonic transducer

2.6 TYPES OF ULTRASONIC WAVES

Basically, there are three types of ultrasonic waves, *viz.,* longitudinal waves, transverse waves and surface waves. Longitudinal waves are also commonly known as compressional waves. Sometimes, these longitudinal waves are also referred to as dilational or irrotational waves.

The second type of wave, *i.e.*, transverse waves, are commonly known as shear waves. Sometimes, these transverse waves are also referred to as distortional waves. The third type of wave, *i.e.*, surface waves are of three types, *viz.*, Rayleigh waves, Lamb waves and Love waves.

This classification of types of ultrasonic waves is based upon the direction of particle vibration when an ultrasonic wave travels through a medium. The vibration of the particles of the medium itself is due to passage of ultrasonic waves through the medium. If the particles of the transmitting medium move in the same direction as the direction of wave propagation, as is shown in Fig. 2.3; the ultrasonic wave is called longitudinal or compressional wave. In case of transverse or shear waves, the particles of the transmitting medium vibrate at right angles to the direction of wave propagation (Fig. 2.4). In isotropic materials, it is possible to transmit both longitudinal and transverse wave.

Surface waves travel over the surface of solid materials (Fig. 2.5). If the surface wave propagates over the surface of a material whose thickness perpendicular to the surface is large compared to the wavelength of the ultrasonic wave, the wave is known as Rayleigh wave. Waves, which are propagated in a solid material and whose thickness is comparable to the wavelength of the ultrasonic wave, are known as Lamb waves. The third type of surface waves, *viz.*, Love waves, travel on the surface without any vertical component. Such waves require a thin layer of some material of different density be present on top of the bulk material below. In case of Lamb waves, the material is in the form of a thin plate. In case of Rayleigh waves, vibrations occur in the plane containing the direction of propagation as well as normal to the surface of the body. However, vibrations normal to the surface are highly damped and there is no particle displacement on the surface perpendicular to the direction of propagation.

2.7 ACOUSTIC IMPEDANCE AND THE NEED FOR COUPLING MEDIUM

Acoustic impedance (Z) of a medium may be defined as the product of the density (ρ) of medium and velocity (v) of sound in that medium, *i.e.*, $Z = \rho v$. When density is expressed in gm/cm^3 and velocity (generally longitudinal velocity) is expressed in cm/sec, unit of impedance becomes gm/cm^2/sec. This is an important parameter to decide, whether or not good transmission of ultrasonic energy shall take place from one medium to another medium. If the density and velocity for the first medium be ρ_1 and v_1 respectively and that for the second medium be ρ_2 and v_2 respectively, then the ratio of the amplitudes of ultrasonic beam in the two mediums is given by the following expression

$$\frac{A_2}{A_1} = \frac{2\rho_1 v_1}{\rho_1 v_1 + \rho_2 v_2} = \frac{2Z_1}{Z_1 + Z_2}$$

This expression clearly demonstrates that for full transmission (*i.e.*, for $A_2 = A_1$) to take place the product $\rho_1 v_1$ should be equal to $\rho_2 v_2$. In other words, the acoustic impedance of the two mediums should be nearly equal. Acoustic impedance of grease, lubricating oils and water etc., is of more or less same order as the acoustic impedance of most of the common solid materials and that is why, grease, oil and water are used as "couplants" (*i.e.*, mediums which provide easy path for the transmission of ultrasonic beam from the transducers to the material under test and the vice versa. Acoustic impedance of air (330 gm/cm^2/sec) is many orders less than that either for oil (1,27,000 gm/cm^2/sec) or for grease (1,49,000 gm/cm^2/sec) and that is why, if air is present in between the ultrasonic transducer and the material under test, A_2 practically becomes zero. Hence, for all ultrasonic testing, it is a must that proper couplant be used. Besides proper couplant, there should be intimate contact between the transducer and the material under test (in case of contact type pulse-echo technique).

2.8 REFLECTION, REFRACTION AND SCATTERING OF ULTRASONIC BEAMS

At the junctions of two materials (*i.e.,* at interfaces), ultrasonic beam undergoes reflection, refraction or scattering and sometimes there exists a combination of these. Defects such as voids, cracks, porosities, inclusions etc., also act as interfaces and, therefore, at such features, there exists reflected, refracted, scattered or a combination of these ultrasonic beam.

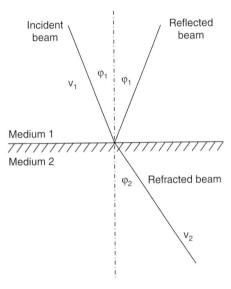

Fig. 2.6 Incident, reflected and refracted beams

Reflection, refraction and scattering of ultrasonic beam follow the same rules as that for optical beams. Figure 2.6 shows incident, reflected and refracted ultrasonic beams. As per the rule of optics, it can be seen from the figure that the angle of incidence is equal to the angle of reflection. Also as given by Snell's law of optics, for ultrasonic beams:

$$\frac{\sin \varphi_1}{v_1} = \frac{\sin \varphi_2}{v_2}$$

The above expression clearly states that if the velocity in the two mediums are different (*i.e.,* $v_1 \neq v_2$) and the ultrasonic beam has an oblique incidence upon the interface (angle of incidence, $\varphi_1 \neq 90°$); the transmitted beam (refracted beam) assumes a new direction of propagation (*i.e.,* not along the same line as that of incident beam) and the new direction is decided by the angle of refraction (φ_2). However, if the velocities v_1 and v_2 in the two mediums are not different (*i.e.,* $v_1 = v_2$), the transmitted beam does not assume a new direction and travels along the same line as that of the incident beam. For such cases, angle of refraction is equal to angle of incidence. (For $v_1 = v_2$, $\sin \varphi_1 = \sin \varphi_2$, or $\varphi_1 = \varphi_2$).

Just as in case of optics, there exists a value of angle of incidence, beyond which there shall not exist any transmitted or reflected beam. This value of angle of incidence, known as critical angle, can be calculated using the following expression:

$$\varphi \geq \sin^{-1}(v_1/v_2)$$
$$[v_2 > v_1]$$

For values of angle of incidence equal to or greater than the critical angle, there does not exist any transmitted or reflected ultrasonic beam and the beam travels along the interface ($\varphi_2 = 90°$).

The ratio of reflected ultrasonic energy and the incident ultrasonic energy is called reflection coefficient and it is usually denoted by letter R. This coefficient can be calculated using the following expression:

$$R = \frac{I_r}{I_i} = \left(\frac{Z_2 - Z_1}{Z_2 + Z_1}\right)^2$$

In the above expression Z_1 and Z_2 are the acoustic impedances of the first and second mediums respectively.

Similarly, the ratio of transmitted ultrasonic energy and the incident ultrasonic energy is called transmission coefficient and is usually denoted by letter T. This coefficient can be calculated using the following expression:

$$T = \frac{I_t}{I_i} = \frac{4Z_1Z_2}{(Z_1 + Z_2)^2}$$

The sum of reflection coefficient R and transmission coefficient T is unity, *i.e.*, R + T = 1.

Besides reflection and refraction, scattering of ultrasonic beam also takes place at the interfaces (or defects); causing loss of ultrasonic energy at these features. This in turn results in reduced amplitude signal of the transmitted and/or reflected ultrasonic beams.

2.9 ULTRASONIC ATTENUATION

Ultrasonic attenuation is a very useful quantity and different materials have different ultrasonic attenuation. A material having higher value of ultrasonic attenuation, attenuates ultrasonic beam to a greater extent as compared to a material having lower value of ultrasonic attenuation. Ultrasonic attenuation is usually expressed in decibels per millimeter (dB/mm). Presence of defects (or discontinuities) in a particular area in a material increases the value of ultrasonic attenuation at that particular area and for areas containing defects, higher the value of ultrasonic attenuation—greater the severity of defects. An increase in the value of ultrasonic attenuation causes the amplitude of transmitted beam to decrease. Hence, by monitoring ultrasonic attenuation one can locate the defects as well as establish their severity. Presence of a flaw is indicated by the fall in the amplitude of transmitted signal on the screen of ultrasonic flaw detector and the amount by which the signal amplitude falls represents the severity of the flaw. How does the decrease in amplitude of transmitted signal relates to increased ultrasonic attenuation, is explained in the next paragraph. It also provides an expression for calculating ultrasonic attenuation in dB/mm.

It is well known that the intensity of transmitted beam (I_t) is related to the intensity of incident beam (I_i) by the following relationship:

$$I_t = I_i\, e^{-\alpha t}$$

where α is a constant and t is the thickness of material through which the ultrasonic beam has travelled. Also, it is well known that intensity is proportional to the square of amplitude, *i.e.,* I $\propto$ A^2. Hence, the aforementioned relationship can be rewritten in terms of the amplitudes of incident and transmitted beams, in the following form:

$$A_t^2 = A_i^2 e^{-\alpha t}$$

or
$$e^{\alpha t} = (A_i / A_t)^2$$

Taking natural logarithm (*i.e.,* $\log_e$ or ln) on both the sides,

$$\alpha t = 2 \ln(A_i/A_t)$$

Changing log to the base e to log to the base 10 (*i.e.,* changing natural logarithm to common logarithm), one would obtain

$$\alpha' t = 2 \log_{10}(A_i/A_t)$$

where α' is another constant to accommodate the conversion constant from natural log to common log. This new constant α' is called attenuation constant or material attenuation and has the unit of Bel. The expression can suitably be modified to obtain attenuation in dB/mm by shifting t (in mm) to the right hand side and changing Bels into dBs (1 Bel = 10 dB). The modified equation becomes:

$$\alpha' \, (\text{dB/mm}) = \frac{20}{t} \, \log_{10} \left(\frac{A_i}{A_t} \right)$$

Hence, by measuring the thickness of the test piece in mm and by noting down the amplitude heights (on the screen of ultrasonic flaw detector) of the incident and transmitted ultrasonic beam, one can calculate the value of ultrasonic attenuation in dB/mm. Sometimes, when the material is highly attenuating, one does not obtain the values of A_i and A_t at the same experimental settings and in such cases one is required to take the help of built-in attenuators or calibrated attenuation switches. This additional attenuation is converted into dB/mm and is then added to the attenuation value obtained using the expression given earlier.

Sometimes, a term known as "total attenuation" is used. This total attenuation is nothing but the sum of "materials attenuation" described above and "surface attenuation" *i.e.,* attenuation due to losses at the interface of material under test and due to the couplant (say water) used for through transmission testing. However, attenuation (or material attenuation) described in the earlier paragraphs is more popularly used as compared to the term total attenuation.

2.10 WORKING OF ULTRASONIC FLAW DETECTORS

There are several types of ultrasonic flaw detectors which are commercially available. Basically, an ultrasonic flaw detector consists of two main units: the ultrasonic signal unit and the oscilloscope unit. The ultrasonic signal unit generates short electric pulses (of duration less than 10 µ sec) which when applied to ultrasonic transducers, provide a short wave train of ultrasonic signal. This section also contains arrangements for receiving ultrasonic signals and converting them into electrical signals for feeding to the oscilloscope unit for proper processing of the received signal and its display on the oscilloscope screen. This section provides sockets for plugging the transmitter and receiver transducers. If the transmitter transducer acts as receiver transducer too, it is plugged to transmitter socket. A toggle switch is provided to select, whether the same transducer is to act as transmitter as well as receiver, or separate transducers for transmitter and receiver. Generally, in pulse-echo technique, the same transducer acts as transmitter as well as receiver; whereas in through transmission ultrasonic testing, separate transducers are used for transmitting and receiving ultrasonic signals. The socket to which transmitter is to be connected, is generally marked "T" and the socket to which the receiver is to be connected is marked "R". The toggle switch is switched to T/R when the same transducer is to be used as transmitter as well as a receiver. It is switched to T-R, when separate transmitter and receiver transducers are used. The ultrasonic section also consists of arrangement for putting in known values of attenuation or a calibrated attenuator. The oscilloscope portion contains the usual controls such as horizontal positioning, vertical positioning, focus, scale illumination, gain, reject, damping, delay etc. Material calibration control, which is a variable potentiometer type control, enables one to adjust the signal transit time display on the oscilloscope screen. The amplitude height of the transmitted or reflected signal on the oscilloscope's screen is adjusted by proper selection of gain, reject and damping settings. For a particular setting of gain, reject and damping; amplitude heights can also be controlled using calibrated attenuation switches. By switching on the attenuation switches, the amplitude falls and by switching off, the amplitude height increases. By selecting proper attenuation switch, one can control the signal amplitude to a desired level.

2.11 INDUSTRIAL APPLICATIONS

Ultrasonic testing is applicable to all type of engineering materials. It can be used to evaluate either metals or plastics or ceramics. It is used for locating practically all type of defects (*i.e.,* surface defects, sub-surface defects, internal defects, metallic and non-metallic inclusions etc.). It is very widely used in the industry for the inspection of incoming materials such as plates, sheets, bars, billets, tubes etc., and for location of voids, cracks etc. Castings (iron, steel, aluminium, brass, copper etc.) having both coarse and fine grain sizes are inspected for blow-holes etc. Ferrous and non-ferrous forging, ferrous and non-ferrous welds, fibre reinforced plastic composites etc. are also inspected for defects using ultrasonic testing technique. Even concrete, rock, wood and other grossly coarse structured materials can be inspected using ultrasonic techniques, but for testing these coarse structured materials, one requires very low frequency (of the order of 50 kHz) ultrasonic transducers as compared to the usual frequency range (1 to 10 MHz) ultrasonic transducers.

Ultrasonic testing finds considerable usage in the field of maintenance inspection too and it is specially so, for locating fatigue cracks. It is used to find out the degree of corrosion which has taken place over a period of time. Ultrasonic testing is also used to measure thickness, particularly when both the sides required for thickness measurement are not accessible resulting in non-adaptability of ordinary means of thickness measurement.

Ultrasonic testing is very extensively used in aerospace industries. It is used for checking airframe components, jet engine rotors, rocket motor casings etc. Ultrasonic testing is extensively used for testing and inspection in railways. It finds marine engineering applications and applications in petroleum refineries too.

Radiography technique is one of the popular industrial non-destructive testing technique and ultrasonic testing technique is usually employed alongwith radiography to complement each other. Generally, speaking, inplaner defects (*i.e.,* defects which lie in a plane parallel to the face) should be detected using ultrasonic technique because these inplaner defects intercept the incident ultrasonic beam to a maximum extent and thereby provide a strong defect signal. Transverse defects (*i.e.,* the defects which lie across the face) should, however, be detected using radiographic technique because transverse defects lie along the path of X-rays (or γ-rays) and thus help in modulating the intensity of transmitted X-rays (or γ-rays) to a maximum extent.

2.12 PULSE-ECHO AND THROUGH TRANSMISSION TESTING

(*a*) **Pulse-echo technique:** This technique is also called reflection technique and it is more widely used method as compared to through transmission method. In pulse-echo method, as the name suggests, a "pulsed" oscillating voltage is applied to a piezo-electric crystal, resulting in an ultrasonic wave train and the "echo" of this ultrasonic wave train is analysed to derive useful information regarding the quality of material under test.

For passing ultrasonic wave train from the ultrasonic transducer to the material under test, a coupling medium is required. The requirement of a coupling medium has already been described in the section 2.8. Common couplants employed have also been mentioned. In general, a coupling medium may be any liquid which will wet the surface and stay between the transducer and the test material during the test. Couplants are also used to fill the uneven spaces and this results in intimate coupling between the transducer and the test material.

There are two types of pulse-echo technique: (*i*) contact type and (*ii*) immersion type. Contact type pulse-echo technique is more popular. Immersion type pulse-echo technique is generally used for automation of testing. The principle of both contact and immersion type pulse-echo technique remains the same. The only difference being that whereas in contact type, the ultrasonic transducer remains in intimate contact with the test material; in immersion type pulse-echo technique, the test material and the ultrasonic transducers do not touch each other but remain dipped in a liquid such as water, oil etc. Scanner assembly for immersion type pulse-echo technique has been described in section 2.13.

For pulse-echo technique, one may either use a single transducer (which acts as transmitter as well as receiver), or a double transducer (or twin transducer) having transmitter in one-half and receiver in the other half, or separate transducers for transmitting and receiving ultrasonic signals. The most popular amongst all the three types mentioned above is the single transducer technique, in which the same ultrasonic transducer acts as transmitter as well receiver (*i.e.*, sends as well as receives ultrasonic signal).

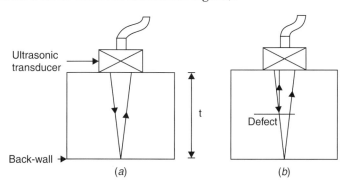

Fig. 2.7 Reflection of ultrasonic beam at interfaces

In pulse-echo technique, a beam of ultrasonic energy is directed into the test material. This beam is intercepted and reflected by the interfaces. The reflected beam is sensed by the ultrasonic transducer (Fig. 2.7) and displayed on the screen of ultrasonic flaw detector (Fig. 2.8). In homogeneous material, ultrasonic beam travels with little loss of energy till the beam is intercepted by discontinuities. If the interface reflecting the ultrasonic beam is a defect, the reflected signal observed on the screen of ultrasonic flaw detector is called a defect echo and if the interface reflecting the ultrasonic beam is the rear surface or the back-wall of the test material, the reflected signal observed on the screen of ultrasonic flaw detector is called back-wall echo. In general, the defect echo and back-wall echo can easily be separated from each other due to the difference in transit time of the two echoes (*i.e.,* more time is taken by the ultrasonic beam to travel to the back-wall and return to the sensing transducer, as compared to the time required for travelling to the defect and returning back to the transducer). For defect free areas, only the initial pulse and back-wall echoes are observed on the screen of ultrasonic flaw detector [Fig. 2.8 (*a*)] whereas in the areas containing defects, in addition to initial pulse and back-wall echoes, there is defect echo too situated inbetween the initial pulse and back-wall echo [Fig. 2.8 (*b*)]. The distance between the initial pulse and back-wall echo represents the thickness of test block (*i.e., x ∝ t*) and, therefore, depending upon the position of defect echo, one can calculate the depth of the defect below the top surface. A defect echo near the initial pulse represents a defect which is near top surface and a defect echo which is near the back-wall echo represents a defect which is near the rear surface.

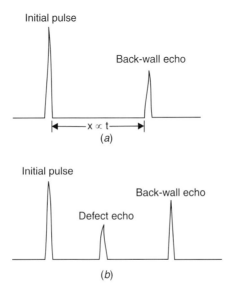

Fig. 2.8 Signal display on the screen of ultrasonic flaw detector

Let us now take a case where the position of defect echo indicates that the defect is at a depth of 100 mm and let us say that this defect provides a defect echo amplitude of 20 mm on the screen of ultrasonic flaw detector. Let us also assume that the specifications require rejection of parts with flaw having a cross-sectional area greater than that of a 2 mm diameter hole. Now to estimate the flaw size, first of all, a test block shall be required. This test block shall be made of same material as that of the parts under test. The test block shall have a 2 mm diameter flat bottom hole approximately 100 mm from the top surface. The reflected signal amplitude from the test block is adjusted using gain control to display an arbitrary amplitude height of, say, 20 mm, 30 mm or 40 mm. This signal height is then compared with the signal height from the flaw/defect in the part under test. If the amplitude of the defect echo at a particular experimental settings is greater than that received from the test block, the cross-sectional area of the defect shall normally be greater than the acceptable size as per the specifications. If the amplitude of the defect echo is less than that received from the test block, the part shall be accepted as per the specifications. One may follow the same procedure, as described above and by using test blocks, containing smaller diameter test holes for establishing the actual size of the defect in the part.

In actual practice, one gets a number of echoes on the screen of ultrasonic flaw detector. These are known as successive back-wall echoes. They occur because the reflected signal from the back-wall, on reaching the top surface, gets reflected again downwards and this signal gets re-reflected from the back-wall providing successive back-wall echoes. The number of successive back-wall echoes, which one would observe on the screen of the ultrasonic flaw detector, depend upon the material of the part under test. For highly attenuating materials, one may have difficulty in even locating one back-wall echo, whereas for little attenuating materials, such as common steels, one may get a number of back-wall echoes. For location of defects, two successive echoes are separated to the maximum possible extent [Fig. 2.8 (a)] and thereafter any defect is detected by the presence of a defect echo in between these two echoes [Fig. 2.8 (b)].

(b) **Through transmission technique:** As is suggested by the name of the technique, "transmission" of ultrasonic energy "through" the material, is the basis of this technique. Different materials absorb ultrasonic energy differently (or say, they attenuate ultrasonic beam differently) and if a bulk material contains defects, the intensity of transmitted ultrasonic beam pertaining to defect-free zone is found to be different as compared to the intensity pertaining to area containing a defect (Fig. 2.9). In a homogeneous material there

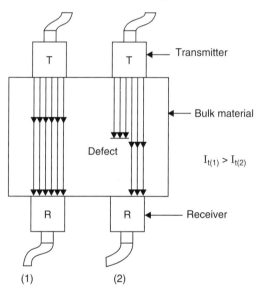

Fig. 2.9 Principle of through transmission ultrasonic testing

is little loss of ultrasonic energy or one may say that good quality homogeneous materials hardly attenuate ultrasonic beam. However, the presence of airpockets (voids, blow-holes, porosites, delaminations, cracks etc.) in the bulk-material change the situation and in areas containing these defects, considerable attenuation of ultrasonic beam takes place. In fact, material containing gross-defects (or materials having highly porous structure) may not allow any transmission of ultrasonic energy, *i.e.*, may attenuate the ultrasonic beam fully. Polyurethane is an example of such materials and for foam, there is no transmission of ultrasonic beam from transmitter to receiver, resulting in zero amplitude received signal on the screen of ultrasonic flaw detector.

For through transmission technique, one requires two separate ultrasonic transducers. (This is unlike pulse-echo technique for which even a single transducer serves the purpose). The ultrasonic transducer, which transmits ultrasonic beam is called "transmitter" and the second ultrasonic transducer, which receives the transmitted ultrasonic signal, after it has travelled through the material, is known as "receiver". Transmitter is kept on one side of the material and the receiver is kept on the other side (Figs. 2.9 and 2.10). The transmitter and receiver are acoustically coupled to the material by using proper couplant. There are two types of through transmission technique: (*i*) contact type and (*ii*) immersion type. Immersion type through transmission technique is more popular. For immersion type through transmission ultrasonic testing, one requires two immersion type ultrasonic transducers. Immersion type ultrasonic transducers have already been described in section 2.5. Immersion type through transmission technique enables one to have a high speed of inspection and automation of scanning operation. It also eliminates the requirement of machining the top and bottom surfaces, which may be required in the case of contact type through transmission testing to obtain intimate contact. Also, it has been observed that in contact type testing if the pressure applied to the hand-held ultrasonic transducer varies, the amplitude of the received signal increases or decreases. These changes in the amplitude not only cause confusion but also make it impossible to record the correct height of the received signal amplitude for calculating the attenuation values. These problems are eliminated by using immersion type testing in which the transducers are not hand-held but are held in their jackets (Fig. 2.10). Immersion type testing should, however, not be used if the immersion of the part would impair the future usefulness of the part.

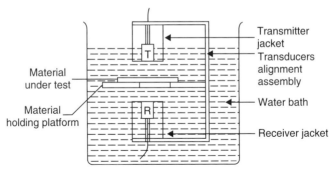

Fig. 2.10 Schematic representation of immersion type through transmission ultrasonic testing

In immersion type through transmission ultrasonic testing, two immersion type ultrasonic transducers (one transmitter and one receiver) are held in a water bath in such a fashion that their central axes are aligned with each other. This is achieved by using properly designed transducers alignment assembly (Fig. 2.10) consisting of two jackets, one for holding the transmitter and the other for holding the receiver. In this aligned position (*i.e.,* when the transmitter and receiver are acoustically coupled to each other by means of some liquid, say water), when practically no attenuation of the incident ultrasonic beam is taking place; the height of the received signal shown on the screen of ultrasonic flaw detector is recorded as A_i (the amplitude of incident ultrasonic beam). By gain control, this height is kept to a maximum to enable observance of the amplitude height of the received signal at a later stage, *i.e.,* when the test object is kept in between the transmitter and receiver transducers. The value of A_i remains the same throughout a particular inspection. At the same experimental settings, now the test material is introduced in between the transmitter and receiver probes. Due to the introduction of test material in the path of transmittance of ultrasonic beam from transmitter to receiver, the amplitude of the received signal, as seen on the screen of ultrasonic flaw detector, falls to a new reading A_t (the amplitude of transmitted ultrasonic beam). From these two readings, A_i and A_t, one may calculate the value of ultrasonic attenuation in dB/mm using the expression given in the section 2.9. If the material is defect-free, the value of A_t and, therefore, the value of ultrasonic attenuation shall remain same throughout the length and breadth of the test material. In other words, during scanning of the test material, a fixed amplitude indicates that the test material is defect-free. If the material does contain a defect, the amplitude of the received signal falls, when the transmitter and receiver are over that particular location, *i.e.*, location having defect. This reduced amplitude value of the received signal (A_t), when used with previously obtained value of A_i, provides a higher value of ultrasonic attenuation. Hence, a fall in the amplitude height of the received signal or an increase in the attenuation value indicates a defect at a particular location. Through transmission technique does not resolve the individual defects and provides only an aggregated effect of all the defects falling in the transmittance path of the ultrasonic beam.

For defects which are very near to the surface (or sub-surface cracks), pulse-echo testing (contact type) fails because of so-called skin-effect. Pulse-echo technique fails for thin-sections too because of the overlapping of defect echoes and back-wall echoes. For a large number of closely spaced defects such as for grossly distributed porosities, individual defect-echoes get superimposed and individual defect resolution is lost, if one uses pulse-echo technique. Hence, for near surface defects, for every thin sections and for grossly distributed defects, immersion type through transmission testing should be adopted.

2.13 SCANNER ASSEMBLIES FOR TRANSMISSION AND PULSE-ECHO TECHNIQUES

Scanner assemblies are required for the automation of ultrasonic inspection. For this purpose, one opts for immersion type testing instead of contact type testing, whether the testing procedure is to be based on through transmission principle or pulse-echo principle. As already mentioned earlier, in immersion testing, the test object along with the ultrasonic transducers are immersed in water or some other suitable liquid couplant. In pulse-echo method, only one immersion type transducer, acting as transmitter and receiver both, is held over the test object (both being submerged in water) (Fig. 2.11). In through transmission testing, two immersion type

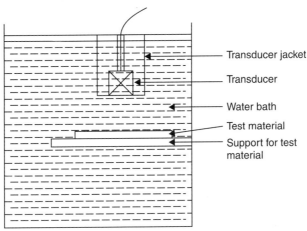

Fig. 2.11 Schematic representation of immersion type pulse-echo ultrasonic testing

transducers are held on the opposite sides of the test object, all being acoustically coupled to each other through water or some other suitable couplant (Fig. 2.10). The equipments used for immersion type testing invariably have the provision of delayed sweep circuits to position and expand or contract the scanning trace on the screen of ultrasonic flaw detector. This delayed sweep arrangement is required because of the fact that the velocity of sound is approximately four times slower in water as compared to the velocity of sound in metallic materials and this results in loss of corresponding amount of viewing area (on the screen) through the extension of base line trace between the outgoing signal from the transducer and the reflection from the top surface of the test object.

For automatic scanning, the transducers are not held manually in position but are placed in suitable jackets for holding them in position. These jackets are themselves attached to a suitable fixture, for either aligning the transmitter and the receiver or to control the angle of incidence. The fixture is then provided with controlled movement along two perpendicular directions (say along X and Y axes) for full scanning of the object. This movement is generally obtained by attaching the fixture to a carriage which could move along the top of the scanning tank and which has the provision of shifting the fixture across too. The scanning assembly looks very much like a typical overhead crane. The carriage could either be moved manually or the carriage could be motorised for obtaining slow speed. By motorising the carriage, the scanning operation becomes quite smooth. The carriage speed should be synchronised with the speed of recording device, say the pen-speed of X-Y recorder. If this motorised carriage is further provided with an indexing device or a transverse positioning device at the end of each longitudinal stroke, one could carry out the ultrasonic inspection of the test material fully without any need of shifting the transducer or the test material manually, during the inspection.

2.14 TYPES OF SCAN

There are three types of scan which are commonly used to present the informations regarding the defects on the screen of ultrasonic flaw detector. These three scans are briefly described below:

A-scan Presentation

This is the most popular type of scan and most of the commercially available ultrasonic flaw detectors use this scan for presentation of information regarding defects in test objects. This type of scan provides information regarding the depth of the defect from the top surface and the amplitudes of initial pulse, the defect echo and the back-wall echo. The position of defect echo on the horizontal base line, on the screen of ultrasonic flaw detector indicates elapsed time (measured from left to right) which is proportional to the depth of the defect from the top surface, as has already been explained in section 2.12. The height of defect echo on the screen of ultrasonic flaw detector (i.e., amplitude of defect echo) indicates the area of cross-section of the defect, as explained earlier. The height of transmitted signal, in case of through transmission ultrasonic testing, is used to calculate the ultrasonic attenuation or to just monitor the location at which the amplitude of transmitted beam falls, indicating the presence of defects. In general, in A-scan presentations, the signal amplitude on the screen of ultrasonic flaw detector represents the intensities of reflected or transmitted beams and can be related to the size of defect, ultrasonic attenuation etc. Figure 2.8 is an A-scan presentation.

B-scan Presentation

Like A-scan presentation, B-scan presentation also provides information regarding the depth of defect. However, instead of providing the information regarding the signal amplitudes, in B-scan, one obtains information regarding defect distribution in cross-sectional view. Therefore, whenever there is a requirement of knowing the shape of defect or distribution of defects, one opts for ultrasonic equipment providing B-scan. Figure 2.12 shows two B-scan indications. The first is that of a big defect and the other is that of small distributed defects. B-scan provides cross-sectional view through the thickness of test block. The top line in a B-scan shows the top-surface of the test block and the bottom line shows the back-wall. In between these two lines, lie the defect signals. Due to the presence of defects, a shadow is created on the trace of back-wall. That is why, the line representing back-wall is shown as broken line and the breaks appear just below the defects (Fig. 2.12). One may expand or contract the B-scan presentation to make it either equal to the actual area being scanned (for easy convertibility of data) or make it larger than the actual size (for easy viewing) or make it smaller than the actual area being scanned (to contain the viewing in the screen area). Generally, after viewing a defect signal on B-scan, reference is made to A-scan too for confirmation and evaluation. For obtaining good B-scan presentation, the size of defect should be large and the transducer used should be a high frequency transducer with a small piezo-electric element.

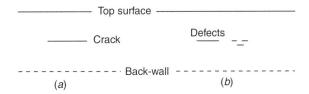

Fig. 2.12 Two B-scan representations (a) Shows one single big defect and (b) Shows distributed small defects

Sometimes, a X-Y recorder is used for image formation of top surface, defect and back-wall. In such applications, the ultrasonic flaw detector with signal output facility is used. Let us say that through transmission immersion type testing is being used and automatic scanning is adopted. Then full scale deflection of recorder pen is adjusted to suit the length and breadth of the test material (*i.e.*, as the scanner assembly scans in X-direction, the recorder pen moves along X-direction and when the scanner assembly shifts in Y-direction, the recorder pen also shifts along Y-direction). The output of ultrasonic flaw detector representing the amplitude height of the received signal is then fed to the X-Y recorder. This way one records the fluctuations in the height of received signal for various locations on the test material (Fig. 2.13). This type of record may be called an ultra-scan and since it provides the details regarding defect distribution in plan view; it may be used in place of B-scan presentation and is actually very similar to C-scan presentation described next.

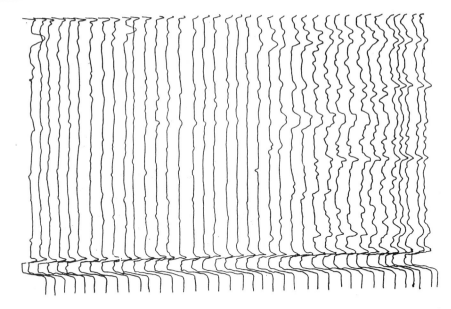

Fig. 2.13 A typical ultrasonic scan obtained using X-Y recorder during through transmission-immersion type ultrasonic testing of carbon fibre reinforced plastic composities

C-scan Presentation

Ultrasonic instruments having C-scan facility show the defect distribution in plan view on the screen of ultrasonic flaw detector. This plan view of defect does not provide any information regarding the depth of defect from the top surface (the information regarding the depth of defect is available in A-scan and B-scan). In C-scan instruments, in addition to the circuit required for B-scan, there is a provision for eliminating unwanted signals such as the initial pulse, interface echo or back-wall echo, which interfere with internal defect signal. Generally, it is achieved by using an electronic gate which allows only those signals which occur within a desired depth range. Due to long time required for scanning a large area, conventional oscilloscopes do not have adequate persistence for practical ultrasonic C-scans. The C-scan display can also be developed with through transmission techniques using appropriate scanning assembly and recording devices, as detailed in the last paragraph of B-scan description.

2.15 SHEAR WAVE APPLICATIONS

Shear wave (or transverse wave) transducers and the characteristics of shear waves have been described earlier in sections 2.5 and 2.6 respectively. Path of shear wave has been shown in Fig. 2.14. Shear wave beam after successive reflections returns to the transducer. The reflection points are called nodes, as shown in Fig. 2.14. Rubbing the material by hand or with a oiled brush along the path of ultrasound interferes with reflections at the nodes. This results in decrease in the amplitude of reflected signal, as obtained on the screen of ultrasonic flaw detector. Thus, by observing the fall in the amplitude of reflected signal, one may locate the point causing reflection. This method is sometimes called "finger damping" or "feeling the location".

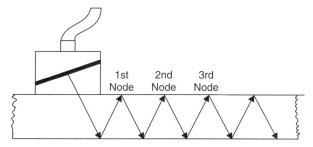

Fig. 2.14 Path of shear wave from a shear wave probe showing different locations of nodes

Shear wave probes are used at places where longitudinal wave probes cannot be used due to one reason or the other such as defect located beneath inaccessible places or rough surfaces etc. It is extensively used for checking weldments for porosity, improper penetration and slag inclusions etc.

2.16 SURFACE WAVE APPLICATIONS

Surface wave transducers and characteristics of surface waves have been described earlier in sections 2.5 and 2.6 respectively. A surface wave transducer transmits surface waves in all directions and these surface waves travel along the surface of the test specimens. These surface waves get reflected at the edges of the specimen (for defect free specimens) or they get reflected at the cracks (or other surface discontinuities). These reflected surface waves produce an echo on the screen of the ultrasonic flaw detector and these echoes are generally very sharp (Fig. 2.15). If one moves the surface wave ultrasonic transducer along the surface of the test specimen, the distance between the transducer and the edge or crack changes and this results in the movement of reflected echo on the screen of ultrasonic flaw detector. Hence, by moving surface wave transducer, one can go on bringing the transmitted and reflected signals together till a time, when the two signals coincide with each other. In this situation, the transducer is just on the top of the crack. If there is no crack on the surface, one will reach the edge of the test specimen to coincide the transmitted and reflected signals. Also, as mentioned in the case of shear waves, one can press his finger on the surface of the test specimen, near the edge of the test

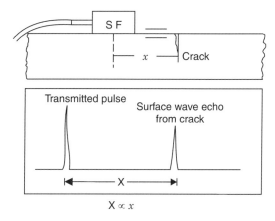

Fig. 2.15 Surface wave application to locate a surface crack

specimen to ascertain, whether or not, the reflected surface wave echo is from the edge. Pressing the surface with finger causes damping of the surface waves and results in the decrease of amplitude (height) of the reflected signal.

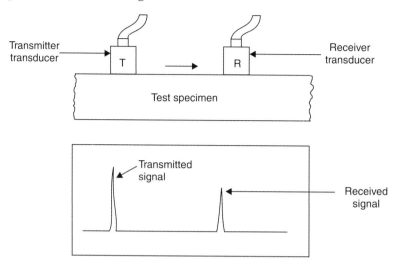

Fig. 2.16 Presence of spurious signal due to surface waves

Sometimes, surface waves provide irrelevant or spurious signals. This is specially so when using separate transducers for transmitting and receiving the ultrasonic beam and when both the transducers are on the same side of the test specimen (Fig. 2.16). In such cases, surface waves are transmitted from transmitter to the receiver along the surface of the test specimen. This results in an echo which may be taken as defect echo if one is not careful. This echo can easily be established as surface wave echo, by moving apart the two transducers because it results in moving apart of transmitted and disturbance signals too. Use of perspex shoes for transducers helps in eliminating these unwanted signals. Sometimes, even when using a single transducer (acting both as transmitter as well as receiver), surface wave echoes provide a spurious defect echo. This takes place due to reflection from the edges of the test specimen and can easily be confirmed by changing the transducer's position on the surface of the test specimen because the movement of the transducer shall result in the movement of this spurious "look alike" of defect echo.

2.17 TYPICAL INDICATIONS

Once an ultrasonic signal indicates the presence of flaw, it is usual practice to cut the test piece and to carry out the microstructural study using a microscope for determining the nature of flaw (*i.e.*, whether the flaw indication is due to presence of a crack, void, inclusion, porosity or flake etc.). Later on, similar ultrasonic indications obtained in similar test pieces are interpreted accordingly. Sometimes, it may be possible to calibrate the ultrasonic signal in terms of the size of flaw, its depth below the surface etc.

Sometimes, one does not go for subsequent microscopic analysis for determining the type of flaw but interprets the signal keeping in view the typical indications obtained from different types of flaws.

Generally, the back-wall echo [Fig. 2.8 (a)] provides a sharp signal and the signal amplitude of the back-wall echo is generally higher than the amplitude of the defect signal [Fig. 2.8 (b)]. A crack which lies parallel to the face of transducer is usually indicated by a sharp indication on the screen of ultrasonic flaw detector; whereas a void or a blow-hole usually provides a "not-so-sharp" type of indication. These "not-so-sharp" indications are also called "bulbous" indications. A crack which is inclined to the surface of transducer may be confirmed by moving the transducer along the surface of the test piece, because inclined cracks cause a lateral shifting of the defect-echo on the screen of ultrasonic flaw detector.

Other typical indications obtained in different type of scans and different types of testing have been provided in earlier sections.

2.18 TEST BLOCKS AND EVALUATING FLAW SIZE

Sometimes to check the proper functioning of ultrasonic probes and ultrasonic flaw detector, one is required to use standard test blocks having known features. These test blocks also help in providing a common basis for comparing results obtained using different instruments, different probes and the results obtained by different fabricators, manufacturers etc. These test blocks act as a basis for standardization.

Fig. 2.17 International Institute of Welders test block (schematic representation)

Test blocks which are most commonly used for standardisation purposes are International Institute of Welders' (I.I.W.) Test Block and ALCOA (Aluminium Company of America) reference blocks. Whereas I.I.W. Test Block contains features such as profile for checking shear wave probes' response, surface features for checking surface wave probes etc.; ALCOA reference blocks consist of a set of eight blocks, each containing a single flat-bottomed hole. Each block having a different diameter hole.

There are many other test blocks used for calibration purposes besides I.I.W. and ALCOA test blocks. Some of these test blocks have conical or spherical holes instead of flat-bottomed hole. Some of the test blocks are slotted at appropriate places to determine the length and width of discontinuities. For estimating flaw size, blocks of different lengths, having flat bottomed hole of different size and length are used. These flat-bottomed holes act as simulated flaws at known depths. These blocks should be made preferably using the same material in which the flaw is to be inspected.

As far as using these test blocks for evaluating flaw size is concerned, one may explain it with an example. Let there be a specification which requires rejection of the parts having flaws with cross-sectional area greater than that of a 2 mm diameter hole. In such a case, first of all a test block is made using the same or similar material. This test block contains a 2 mm diameter flat bottomed hole approximately 100 mm from the back face. Now, using this test block, the signal response is adjusted (by gain control) to display an arbitrary signal amplitude of say 25 mm (The choice of signal amplitude is actually immaterial). This signal amplitude is now compared with the signal amplitude from the actual flaw in the part to be inspected. If the signal amplitude (at the same setting) is more than that received from the test block, the cross-sectional area of the flaw will normally be greater than the acceptable limit. If the signal amplitude is shorter, the test object may be accepted. The same test procedure can be repeated using reference blocks having smaller diameter test holes (*e.g.*, 1.5 mm, 1.25 mm, 1 mm etc.) to evaluate the actual flaw size.

2.19 RESONANCE TECHNIQUE

Just like pulse-echo technique and through transmission technique, resonance technique is also one of the ultrasonic techniques to evaluate flaws in various materials. Basically, in resonance technique, one requires to have a tunable variable frequency continuous wave oscillator, which is used to drive a transducer. As the oscillator is tuned through its tuning range, the specimen vibrates in resonance as soon as the frequency of the oscillator reaches the resonant frequency for that particular thickness of the specimen, *i.e.*, whenever thickness of the specimen is equal to an integral number of half wavelengths of the ultrasonic wave. Standing wave pattern in solid materials at which resonance occurs is shown in Fig. 2.18.

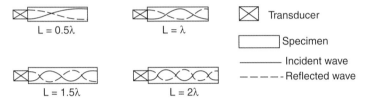

Fig. 2.18 Different specimen lengths for resonance to occur

While detecting flaws using resonance technique, one has to only look for the presence or absence of thickness indication. Thickness resonance is indicated on the screen of ultrasonic instrument by the presence of a large number of resonance peaks. Resonance technique may be used for evaluating flaws in test objects which have their opposite sides smooth and parallel (such as plates, sheets, blocks, bars etc.). In this technique, first of all the ultrasonic transducer is acoustically coupled to one of the parallel surfaces and by using the oscillator; resonant frequency is obtained for the thickness of the test object. At this point, the oscilloscope screen of the ultrasonic instrument shows a number of resonance or harmonic peaks (or in other words, thickness indication is provided on the screen). This in turn indicates that there are no voids, cavities or porosities covering any appreciable area under the ultrasonic transducer. For flaws, which are not parallel to the surface, thickness indication either disappears completely or the indication gets reduced considerably. Hence, by watching for significant changes in the resonance peaks, one may evaluate flaws in the test objects. For flaws which are parallel to the surface, the resonance peaks for the total thickness, do not appear and they appear for the reduced thickness, thereby enabling one to determine the depth of the flaw. For flaws which are perpendicular to the surface, appreciable changes in thickness indication are obtained only for flaws having width comparable to the crystal size. For small size perpendicular flaws, thickness indication changes very little.

Resonance technique is also employed for non-destructive evaluation of certain types of bonds. The bonds having adhesive areas (*i.e.*, areas of actual separation) are easily detected using resonance technique. Resonance technique is also used to test bonds between sheet metals and rubbers. Whereas, for good bonds, the thickness indications are either absent or very small; for poorly bonded areas, good thickness indication are obtained. For applications, where there is excessive variation in thickness (*e.g.*, as in the case of severe corrosion of steel plates), resonance technique cannot be put to use. Resonance technique finds widest application in the area of thickness measurement, especially when there is only one side accessible. The resonance thickness measurement technique has been described in the section to follow.

2.20 USE OF ULTRASONICS FOR THICKNESS MEASUREMENT

Use of ultrasonic technique for thickness measurement is quite common these days. Specially designed ultrasonic instruments, designed solely for thickness measurement, are commercially available and are known as Ultrasonic Thickness Gauge.

Resonance thickness measurement is one of the commonest thickness measurement technique. As described in the previous section, thickness resonance occurs whenever the thickness of the material is equal to an integral number of half wavelengths of the ultrasonic wave. This has also been shown in Fig. 2.18. In this figure the points of maximum displacement are called antinodes and the points of minimum displacement are called nodes. As can be observed from Fig. 2.18, for resonance to occur, the distance between two successive nodes or successive antinodes should be equal to a half wavelength. Also shown in figure, are the thickness values in terms of wavelength (λ) for the specimens, *i.e.*, as t is equal to 0.5 λ, λ, 1.5 λ, and 2 λ respectively. Using the well known expression between velocity, frequency, and wavelength (*i.e.*, $v = n\lambda$); one may write the expression for the fundamental frequency at which thickness resonance shall occur. Hence,

$$n_1 = v/\lambda = v/2t$$

similarly $$n_2 = v/t, n_3 = 3v/2t \text{ and } n_4 = 2v/t$$

or $$n_2 = 2n_1, n_3 = 3n_1, \text{ and } n_4 = 4n_1$$

These frequencies n_2, n_3 etc. are known as harmonics or multiples of the fundamental frequency. It can be noticed that the frequency difference between any two adjacent harmonics is equal to the fundamental frequency. Hence, one may use the following expression for evaluating the thickness of a part, after determining two adjacent harmonic frequencies.

$$t = \frac{v}{2(n_n - n_{n-1})}$$

Pulse-echo technique can also be used for thickness measurement. While employing pulse-echo technique for inspecting a material for flaws etc., one obtains successive back-wall echoes. Thus, distance between two successive back-wall echoes on the abscissa of ultrasonic flaw detector's screen represents twice the thickness of specimen being inspected. Thus, there exists a proportional relationship between the distances between successive echoes and the specimen's thickness. Hence, it is always possible to calibrate abscissa's divisions in terms of thickness. This calibration holds good for a particular setting of the ultrasonic instrument and for that particular material. Standard step blocks can be made of different materials for calibration purposes. Generally, the commercially available step blocks for thickness measurements have eight steps for 1 cm to 8 cm in steps of 1 cm. Once the distance between two successive echoes has been calibrated, thickness measurements can easily be carried out by just looking at the positions of two successive back-wall echoes on the screen of the ultrasonic flaw detector.

2.21 A FEW APPLICATIONS OF ULTRASONICS IN MEDICAL SCIENCES

Non-destructive testing techniques play a very significant role in the field of medical sciences too. Whereas in the field of engineering, NDT techniques are used for evaluating defects in engineering materials and for studying their physical properties; NDT techniques are used in the field of medical sciences for diagnostic purposes and for therapeutic uses. In the previous Chapter, a mention was made as to how wide is the use of X-ray technique of non-destructive testing in the field of medical sciences. One hears about medical X-rays and Computerized Axial Topography (CAT) – which is a three-dimensional radiography, scans very frequently. Just like X-rays, ultrasonics are also used very widely in the field of medical sciences. Ultrasonic foetal detector is widely used by obstetricians. Ultrasonography and ultrasonic CAT scanner are finding increasing application for diagnostic purposes and radiologists are readily adopting to these methods. Ultrasonic therapy units are also in popular use at therapeutic centers.

Bio-medical engineers have developed ultrasonic technique for removal of kidney stones without surgery (shock-wave lithotripsy). In this technique, exact position of kidney stone is located by taking X-ray radiographs from two perpendicular directions. Once the location of the stone is known, three or more focused ultrasonic probes are arranged around the body (externally) in such a way that their focal points coincide with the position of the stone within the body. Water-jacket around the body is used for transmitting ultrasonic pulses from the focused transducers to the body and thereafter the body itself acts as transmitting medium. These ultrasonic waves of definite power and frequency act as shock-wave and crush the kidney stone. The crushed stone (in the form of little pieces) is then removed from the body while the patient urinates.

It is believed that ultrasonic beams are not associated with common radiation hazards and, therefore, ultrasonic techniques for medical usage are being developed vigorously.

2.22 DETERMINATION OF GRAIN SIZE USING ULTRASONICS

Through transmission ultrasonic testing can be used for determining grain size. For this purpose, one may obtain a calibration curve relating ultrasonic attenuation (dB/mm) to grain size (mm). As an example, for obtaining the calibration curve for brass, one may prepare brass specimens (40 mm diameter × 10 mm long) and heat-treat them at different temperatures so as to obtain average grain diameter ranging from 0.025 to 0.15 mm. Now the heat treated brass samples may be put in a water tank (one by one) in between transmitter and receiver probes, as has already been described while describing through transmission ultrasonic testing. The amplitude of received ultrasonic wave may be noted (for each sample) with and without brass samples in between the transmitter and receiver probes. These amplitude readings will enable one to calculate attenuation (dB/mm) of different brass samples by using the expression given in section 2.9. When attenuation of different brass samples is plotted against the average grain size of these individual samples, one obtains the calibration curve. For measuring grain size, one needs to adopt ordinary polishing and micrographic test facilities and use of standard grain size references. It shall be observed from this calibration curve that the attenuation (dB/mm) of samples increases with increasing grain size. These tests can be conducted using immersion type ultrasonic probes having frequency 5 MHz or so. Once the calibration curve is obtained, one may evaluate the grain size of brass samples with unknown grain size, by just measuring its ultrasonic attenuation (dB/mm), nondestructively.

CHAPTER 3

Eddy-Current Testing

Eddy-current testing is based on well known phenomenon of electromagnetic induction and is applicable for non-destructive evaluation of all electrically conducting materials including electrically conducting Carbon Fibre Reinforced Plastics (CFRP) having fibre volume fraction 40% and higher. The technique however cannot be used for non-destructive evaluation of electrically non-conducting FRP composites such as glass-fibre reinforced plastics.

3.1 ELECTRICAL PROPERTIES OF CARBON FIBRE REINFORCED PLASTICS

Carbon fibres are good electrical conductors and the electrical resistivity of carbon fibres is found to be very low (of the order of $\mu\Omega$). When these carbon fibres are used for reinforcing plastic matrix to obtain unidirectional CFRP composites with practical fibre content (*i.e.*, 40% volume fraction and higher), there exists a great deal of fibre to fibre contact and consequently such composites too become good electrical conductors, both parallel and transverse to fibre direction. Typically at 55% fibre volume fraction, longitudinal resistivity is about 8×10^{-5} Ωm and transverse resistivity is 4×10^{-3} Ωm. Complex lay-ups and chopped fibre composites can have even more fibre to fibre contact resulting in lower values of transverse resistivity.

CFRP possesses a high degree of electrical anisotropy. Longitudinal conductivity of unidirectional CFRP composite is controlled by electrical properties of carbon fibres and transverse conductivity is controlled by electrical properties of matrix material (*e.g.*, that of epoxy resin in case of carbon fibre reinforced epoxy composites). If one considers the anisotropy ratio of electrical conductivity for an ideal unidirectional CFRP composite (an ideal unidirectional CFRP composite is one in which all the fibres are straight and parallel and are fully wetted by resin and are situated at regular intervals throughout the matrix), it will be infinity. The value of anisotropy ratio of electrical conductivity for CFRP composite having 55% fibre volume fraction (commonly used CFRP composite) is, however, of the order of 50 or so.

Longitudinal conductivity of unidirectional CFRP composites varies linearly with the fibre volume fraction. The angular dependence of electrical properties of CFRP has been studied and it has been reported that just like for isotropic conducting materials, for unidirectional CFRP too, electrical conductivity and electrical resistivity have inverse relationship along principal directions. In general direction, however, this reciprocality is not valid. It is reported in the literature that while measuring electrical resistivity of CFRP, proper electrical contacts are not easily made and accuracy of resistivity measurement is therefore limited to ±7%.

3.2 PRINCIPLE OF EDDY-CURRENT TESTING

When a coil carrying alternating current is brought near an electrically conducting material, eddy-currents are induced in the material by electromagnetic induction (Fig. 3.1). Magnitude of induced eddy-currents depends upon the magnitude and frequency of alternating current; distance between current carrying coil and material under test; presence of defects or inhomogeneities in material and physical properties of material. The induced eddy-currents modulate the impedance of the exciting coil or any secondary coil situated in the vicinity of test material. The difference between original coil impedance and modulated coil impedance (due to presence of eddy-current) is monitored to obtain meaningful information regarding presence of defects or changes in physical, chemical or microstructural properties.

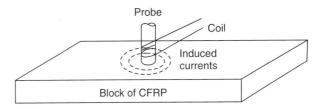

Fig. 3.1 Eddy-current probe over a block of electrically conducting material

As mentioned in the earlier paragraph, eddy-current response depends upon distance between test coil and specimen. Even a slight variation in distance between test coil and specimen may change eddy-current response signal significantly and may wrongly be interpreted for variations in structure of test material. To eliminate such a possibility, it is always advisable to design the probe in such a way that during the test, the eddy-current probe remains always in contact with the test material. Sticking a small acrylic piece to spring mounted magnetic core of test coil helps in obtaining and maintaining proper contact between probe and test material.

3.3 APPLICATION OF EDDY-CURRENT TESTING

As far as metals are concerned, eddy-current technique has been used for quite some time for varied applications such as detection of cracks, porosities and inclusions, metal sorting, evaluation of plate or tubing thickness, measurement of coating thickness and thickness of non-conducting films on metallic bases etc.

For electrically conducting CFRP composites, eddy-current technique can usefully be employed for quantification and location of defects and other inhomogeneities. It can also be used for evaluation of fibre volume fraction in unidirectional CFRP laminates and lay-up order in crosspiled CFRP laminates.

Eddy-currents are generally concentrated near the surface of conducting material and therefore eddy-current technique is most effective for locating near surface/sub-surface irregularities. The magnitude of eddy-current falls-off exponentially as the depth below the surface increases. The depth upto which eddy-current can penetrate in any appreciable amount in any electrically conducting material is called depth of peneration (d) and it can be calculated using following expression:

$$d = (\pi f \mu \sigma)^{-0.5}$$

where $\quad d$ = depth of peneration in metre

f = frequency of alternating current in Hertz

μ = magnetic permeability (if the material is non-magnetic such as CFRP, it is taken as 4×10^{-7} Henry/metre), and

σ = electrical conductivity (mhos/metre)

For CFRP, which is usually used in laminate form, depth of penetration is never a serious problem. In case of CFRP laminates, one can easily and quickly ascertain adequate depth of peneration by moving a metallic piece or tip of a screw driver or alike below CFRP laminate (probe being on top surface side) and observing any change in response of eddy-current circuit frequency or related d.c. voltage. If the movement is associated with a change in eddy-current response signal, it indicates that eddy-current is penetrating right through the thickness of CFRP laminate under test.

3.4 EDDY-CURRENT PATH

For eddy-current probes, which are magnetic core type, shape of induced eddy-current path depends upon geometry of the core. For circular core probes, (which can be made by taking a circular ferrite rod and by winding 4 to 6 turns of 20 SWG enamel coated copper wire around the core), induced eddy-current path in all electrically conducting isotropic and homogeneous material is found to be circular. However, if one uses a circular probe for eddy-current testing of unidirectional CFRP composites, which are orthotropic, induced eddy-current path is found to be elliptical. Major axis of this elliptical path lies along fibre direction and length of this major axis is found to be roughly three times the diameter of the circular core of eddy-current probe. This elliptical eddy-current path gets distorted at locations where there exists any defect or inhomogeneity. This distortion of eddy-current path alters the impedance of exciting or pick-up coil. Thus, by monitoring changes in coil impedance, one can locate defects and other inhomogeneities.

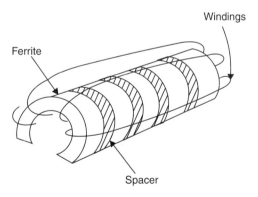

Fig. 3.2 Horse shoe eddy-current probe

For non-circular probes such as horse-shoe probe (Fig. 3.2), induced eddy-current path too is found to be non-circular even in isotropic and homogeneous electrically conducting materials. Non-circular eddy-current path gets affected by relative rotation of probe and the material under test. The changes in the induced eddy-current path due to the relative rotation of probe, causes impedance of probe to vary. The variation in impedance depends upon the shape and the size of the probe and on the anisotropy of the material under test. Thus, by using a standard size and standard shape probe, one may study the material anisotropy.

3.5 EDDY-CURRENT COILS/PROBES

Shape and size of eddy-current exciting and pick-up coils vary greatly depending upon specific applications. Some applications may require a single turn winding whereas others may require several windings. Also, coils may be of air-core type or they may have a magnetic core (such as ferrite rod or dust-iron bar) for obtaining higher sensitivity or resolution. Sometimes, proper shielding is also provided to test coil for increased resolution.

Exciting coil and pick-up coil can be of two types. They can either be resonant type or driven type. Resonant type test coils are also known as absolute coils. While using absolute coils, measurements are made without a direct reference to or for comparison purposes. For good specimens, voltage output from the two coils is found to be zero and for specimens having different structures than standard, an output voltage proportional to difference in quality, is obtained. Differential type test coils are, however, not very sensitive to gradual changes in structural properties and therefore, in practice absolute test coils are found to be far more popular.

3.6 EDGE-EFFECT

Induced eddy-currents get affected by the proximity of edges of sample and therefore eddy-current signals near the edges of a sample under test may be quite misleading. In case of unidirectional CFRP laminates, as mentioned earlier, eddy-current probe which has say 4 mm diameter ferrite core, one should avoid interpreting signals obtained from around the periphery of CFRP laminate, say 6 mm from the edges of CFRP laminate.

3.7 RECENT TRENDS IN EDDY-CURRENT TESTING

Eddy-current technique has improved significantly to satisfy the stringent quality requirements of present day high technology industries. Considerable betterment in respect of both the sensitivity and speed of testing has been achieved. Microprocessor based systems are finding extensive applications for on-line and in-service inspection. These developments are, however, limited to specific applications and equipments. There exists a need to diversify these developments to other areas too.

3.8 HIGH FREQUENCY EDDY-CURRENT TEST

In certain applications where one has to detect only surface cracks which are very fine, *e.g.,* fatigue cracks in metallic parts; one has to go for higher test frequency. As one would observe from the expression for calculating the depth of penetration (given earlier), depth (d) decreases with increasing frequency (f). Higher test frequency results in higher current density on test surface resulting in better resolution of surface cracks. In case of CFRP laminates, increase in test frequency also results in modified eddy-current path. Eddy-current path at lower test frequencies (less than 15 MHz) and modified eddy-current path at higher test frequencies (greater than 15 MHz) for unidirectional CFRP laminates, are shown in Figures 3.3 and 3.4.

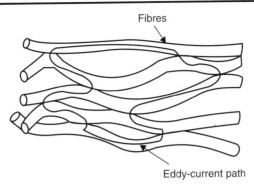

Fig. 3.3 Resistive eddy-current path

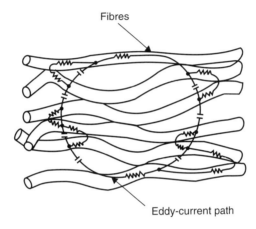

Fig. 3.4 Resistive reactive eddy-current path

3.9 ELECTRICAL ANALOGUE OF EDDY-CURRENT TEST

Eddy-current path in unidirectional CFRP is partially resistive (along fibres) and partially capacitive (across resin, jumping between fibres). At lower test frequencies, reactance of the capacitive path becomes appreciably high and current path across the resin matrix becomes an unfavourable path. Consequently, eddy-currents flow along fibres and the current flows from one fibre to another fibre at the points of fibre/fibre contact, is shown in Fig. 3.3. This path is known as resistive path and eddy-current circuit based on electrical analogue of resistive path is used for the detection of cracks in CFRP laminates.

At frequencies in excess of 15 MHz, the interfibre reactance (*i.e.*, reactance across resin matrix) becomes comparable to the resistance of the alternative fibre condition path. At very high frequencies, the interfibre reactance becomes quite low as compared to the resistance of the alternative conduction path. Consequently, an eddy-current path which is resistive-capacitive and shown in Fig. 3.4 becomes more appropriate. This eddy-current path is partially resistive (along fibres) and partially capacitive (jumping between fibres) shunted by a longer resistive path (due to fibre/fibre contact).

Eddy-current circuit possess inductance by virtue of its area and, therefore, conventional electrical analogue of eddy-current testing (such as that for metallic materials) is represented by two coupled inductors (representing the probe and the current loop in the material) with a resistor in series with the second inductor (Fig. 3.5).

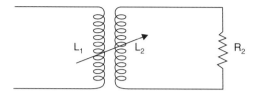

Fig. 3.5 Induced current circuit

3.10 THEORETICAL ANALYSIS OF EDDY-CURRENT CIRCUIT

Eddy-current circuit is a tuned circuit and it is more or less damped depending on the values of the resistive component. In the series circuit there are two resistances shown, *viz.*, R_2 and R_s (Fig. 3.6). R_2 represents fibre resistance and R_s represents shunt resistance across

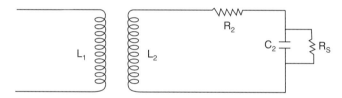

Fig. 3.6 Electrical analogue of eddy-current probe and CFRP laminate

the capacitance C_2. R_2 and R_s are of similar magnitude and consequently R_s plays a dominant role in determining the frequency response of the circuit. The analysis of the circuit is described below:

As is well known, the primary impedance Z_p in the presence of a secondary is given by

$$Z_p = j\omega L_1 + \frac{\omega_2 M_2}{Z_s} \qquad \qquad ...(1)$$

where ω is the working frequency,

M is the mutual inductance between L_1 and L_2 and

Z_s is the series impedance of the secondary.

$$Z_s = j\omega L_2 + R_2 + \frac{1}{1/R_s + j\omega C_2}$$

$$= \frac{1}{1 + \omega^2 R_s^2 C_2^2} \{R_2 + R_s + \omega^2 R_2 (R_s C_2)^2$$

$$+ j\omega [L_2 + \omega^2 L_2 (R_s C_2)^2 - R_s (R_s C_2)]\}$$

Substituting into (1) and putting

$$R_2/L_2 = \alpha,\ R_s/L_2 = \beta,\ R_s C_2 = \gamma,$$

$$M^2 = K^2 L_1 L_2$$

gives the following as the changes in the primary probe impedance produced by the presence of the test piece.

(*a*) an addition in resistive component

$$\Delta R = \frac{\omega L_1 \omega k^2 (1 + \omega^2 \gamma^2) [\alpha(1 + \omega^2 \gamma^2) + \beta]}{[\alpha(1 + \omega^2 \gamma^2) + \beta]^2 + \omega^2 [1 + \omega^2 \gamma^2 - \beta \gamma]^2} \qquad \ldots(2)$$

(*b*) a changing in the reactance of the primary by

$$\Delta X = \frac{-\omega L_1 \omega k^2 (1 + \omega^2 \gamma^2) \omega (1 + \omega^2 \gamma^2 - \beta \gamma)}{[\alpha(1 + \omega^2 \gamma^2) + \beta]^2 + \omega^2 [1 + \omega^2 \gamma^2 - \beta \gamma]^2} \qquad \ldots(3)$$

In order to get a picture of these equations, it is necessary to simplify them using approximations.

Approximation 1

Experiment shows that there is a frequency at which the change in reactance is zero. The only term in equation (3), which can give this behaviour is $1 + \omega^2\gamma^2 - \beta\gamma$. If ω_0 is the frequency at which $\Delta X = 0$.

$$\omega_0^2 = \frac{\beta\gamma - 1}{\gamma^2} = \frac{\beta}{\gamma} - \frac{1}{\gamma^2} = \frac{1}{L_2 C_2} - \frac{1}{(R_s C_2)^2}$$

ω_0 is usually in the region of 10 MHz. If CFRP specimens having a wide range of volume fractions are tested, it would be unlikely that C_2 and R_s would be very repeatable and, therefore, it seems that $1/(R_s C_2)^2$ is a correction factor to a much larger $1/L_2 C_2$.

Taking this correction to be less than 50%

$$\beta/\gamma > 2\frac{1}{\gamma^2}$$

or

$$\beta\gamma > 2$$

as

$$\omega_0^2 \gamma^2 = \beta\gamma - 1$$
$$\omega_0{}^2\gamma^2 > 1$$

Approximation 2

In practice, reasonable resolution requires a working frequency approaching 100 MHz, *i.e.*,

$$\omega = 10\omega_0$$

$$\therefore \qquad \omega^2\gamma^2 >> 1 \qquad \ldots(4)$$

at normal working frequencies.

Approximation 3

$$\omega^2\gamma^2 - \beta\gamma = \omega^2 R_s{}^2 C_2{}^2 - (R_s/L_2) \cdot R_s C_2$$
$$= \omega R_s{}^2 C_2 [\omega C_2 - 1/\omega L_2]$$

But

$$\omega_0 C_2 = 1/\omega_0 L_2 \quad \text{and} \quad \omega > \omega_0$$

$$\therefore \qquad \omega^2\gamma^2 - \beta\gamma = \omega R_s{}^2 \omega C_2{}^2$$
$$= \omega^2\gamma^2 \qquad \ldots(5)$$

Approximation 4

If we assume that the properties of the system are entirely determined by the shunt resistance R_s, *i.e.*, $R_2 = 0$.

Then $\alpha = 0$. From equation (2), making use of approximation 2 (equation 4), we have

$$\Delta R = \frac{\omega L_1 . \omega^2 \gamma^2 . \beta}{\beta^2 + \omega^2 (\omega^2 \gamma^2 \beta \gamma)^2}$$

As the working frequency ω is well above ω_0, the natural resonant frequency of the secondary, approximation 3 (equation 5) applies

$$\therefore \qquad \Delta R = \frac{L_1 k^2}{\omega^2 \gamma^2}$$

$$= k^2 \frac{L_1}{L_2} . R_s . \frac{1}{\omega^2 (R_s C_2)^2} \qquad \qquad ...(6)$$

If the material is a metal, the result would be

$$\Delta R = K^2 . \frac{L_1}{L_2} . R_s$$

i.e., the presence of the inter-fibre capacitance, therefore, appears to reduce the reflected resistance.

From equation (3) and again making use of approximation 2 (equation 4)

$$\Delta X = - k^2 . \omega L_1 \qquad \qquad ...(7)$$

which is the same result as would be obtained if the material under test were a metal. Work of Owston shows that even with $\omega = 10\omega_0$, ΔX is less than 25% of that when the sample was metal. It, therefore, appears that it is not valid to assume that $R_2 = 0$.

Approximation 5

If we suppose that R_s and R_2 are both finite and both affect the performance of the system, *i.e.*, $\alpha = R_s/L_2$ is of similar order to $\beta = R_2/L_2$. Approximations (2) and (3) (equations 4 and 5) are still applicable, therefore from equation (2)

$$\Delta R = \frac{\omega L_1 . \omega k^2 . \omega^2 \gamma^2 . [\alpha \gamma^2 \gamma^2 + \beta]}{\alpha (\omega^2 \gamma^2 + \beta)^2 + \omega^2 (\omega^2 \gamma^2)^2} \qquad \qquad ...(8)$$

Similarly, from equation (3)

$$\Delta X = \frac{\omega L_1 . \omega k^2 . \omega^2 \gamma^2 . \omega^2 \gamma^2}{(\alpha \omega^2 \gamma^2 + \beta)^2 + \omega^2 (\omega^2 \gamma^2)^2} \qquad \qquad ...(9)$$

It is attractive to assume that as $\omega^2 \gamma^2 > 1$ and α and β may be of the same order, $\alpha \omega^2 \gamma^2 \beta$ and equations (8) and (9) can be approximated as

$$\Delta R = \frac{\omega L_1 \omega k^2 a}{\alpha^2 + \omega^2}$$

$$= k^2 \frac{L_1}{L_2} R_2 \frac{1}{(R_2/\omega L_2)^2 + 1} \qquad \qquad ...(10)$$

and $\qquad \qquad \Delta X = - k^2 \omega L_1 . \frac{1}{(R_2/\omega L_2)^2 + 1} \qquad \qquad ...(11)$

The reactance term now has the required form, being less than for a metal sample by a factor of

$$\frac{1}{(R_2/\omega L_2)^2 + 1}$$

As the factor is commonly 1/5 or less (*i.e.*, 20%), it follows that:

$$(R_2/\omega L_2)^2 = 4 \qquad \qquad ...(12)$$

Therefore, approximately

$$\Delta X = -k^2.\omega L_1.(\omega L_2/R_2)^2 \qquad \qquad ...(13)$$

From the basis of the tuned circuit analogue, R_2 represents the electrical resistance of the fibres to conduction along the fibre direction. For the composite as a whole R_2 is proportional to $1/V_f$ where V_f is the fibre volume fraction. Hence, equation (13) becomes

$$\Delta X \text{ proportional to} - (V_f)^{-2} \qquad \qquad ...(14)$$

Using the same approximation, equation (10) becomes

$$\Delta R = k^2 \frac{L_1}{L_2} \frac{(\omega L_2)^2}{R_2} \qquad \qquad ...(15)$$

i.e., $\qquad \qquad \Delta R$ is proportional to $V_f \qquad \qquad ...(16)$

Hence, both ΔR and ΔX should depend on V_f which is not in accordance with the experimental observations.

Equations (10) and (11) are of the form

$$\Delta R = A \frac{x}{\alpha^2 x^2 + 1}$$

and $\qquad \qquad \qquad \Delta X = B \dfrac{1}{\alpha^2 x^2 + 1}$

respectively, where A, B and α are constants as far as variations in volume fraction are concerned. Hence,

$$\frac{d}{dx}(\Delta R) = A \cdot \frac{a^2 x^2 - 1}{(a^2 x^2 + 1)^2} \qquad \qquad ...(17)$$

$$\frac{d}{dx}(\Delta X) = -B \cdot \frac{1}{x} \frac{2\alpha^2 x^2}{(\alpha^2 x^2 + 1)^2} \qquad \qquad ...(18)$$

Hence, the fractional changes in the impedance parameters are related to the fractional change in x by

$$\frac{d(\Delta R)}{\Delta R} = \frac{a^2 x^2 - 1}{(a^2 x^2 + 1)} \frac{dx}{x}$$

$$\frac{d(\Delta X)}{\Delta X} = \frac{-2a^2 x^2}{(a^2 x^2 + 1)} \frac{dx}{x} \qquad \qquad ...(19)$$

If the effect of a piece of CFRP placed near the probe were to change the reactance by half the change produced by a metal; from equation (11), we would have

$$\alpha^2 x^2 + 1 = 2$$

$\therefore \qquad \qquad \qquad \alpha^2 x^2 = 1$

and substituting in (19)

$$\frac{d(\Delta R)}{\Delta R} = 0; \quad \frac{d(\Delta X)}{\Delta X} = -\frac{dx}{x}$$

i.e., the resistance component of probe impedance would not show any changes with small changes in volume fraction. For values of $a^2x^2 + 1 \neq 2$ the resistive component will show some changes but even for $a^2x^2 + 1 = 5$.

$$\frac{d(\Delta R)}{\Delta R} = 0.6\frac{dx}{x}; \quad \frac{d(\Delta X)}{\Delta X} = 1.6\frac{dx}{x} \qquad \qquad ...(20)$$

i.e., the reactive component remains more sensitive to changes in volume fraction.

This is in agreement with experimental observations described later and it confirms that equipment designed to study variations in volume fraction must follow the reactive component of probe impedance.

The above analysis can be summarized as:

(*i*) Variations in fibre volume fraction may not affect the resistive component of probe impedance.

(*ii*) Variations in fibre volume fraction always appear to affect the reactance of the probe over the normal working frequently range.

(*iii*) To avoid complex dependence on fibre volume fraction, the test frequency should be as high as possible.

3.11 FIBRE VOLUME FRACTION MEASUREMENT BY EDDY-CURRENT TECHNIQUE

Based on the findings of the theoretical analysis of eddy-current circuit, a system was designed. The system monitored the reactance of the probe to determine the variation in the fibre volume fraction. The probe was made the inductance of the tank circuit of a free running oscillator (Fig. 3.7). Feedback on TR_1 maintained the oscillation, while TR_2 an emitter follower gave an impedance match of coax cable to the measuring instrument. As the reactance of the probe changed with the changes in the specimens presented to it, so did the frequency of the oscillator.

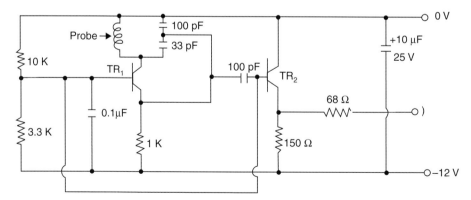

Fig. 3.7 Reactance sensitive eddy-current circuit

The output from the oscillator was fed to a frequency deviation meter. This deviation meter measured the static deviations from a preset mean frequency and it was capable of recording 0.2 MHz deviations while working at frequencies of the order of 50 MHz.

The output of the frequency deviation meter was fed to a system which combined this signal with the position transducers' signals of a scanner assembly on which the CFRP laminates under test were mounted. This combined signal was then fed to X-Y plotter to get a X-Y display of the variation in the frequency deviation meter's reading, as the probe scanned the CFRP laminate. This display is known as "eddy-scan" of the laminate (Fig. 3.8).

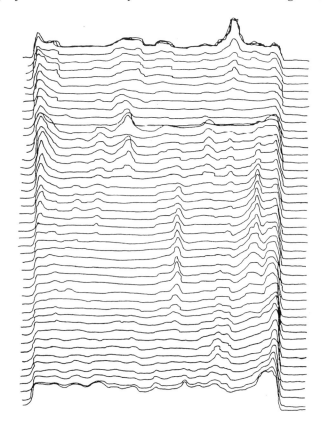

Fig. 3.8 Eddy-scan of a CFRP laminate

The probe used with this reactance sensitive eddy-current circuit, was a hexagonal dust iron rod, 3 mm across faces and 10 mm long. The rod was wound with 4 turns of 20 SWG enamel coated copper wire. A small piece of acrylic was put at the end of the rod and the probe was designed so as to keep the probe and CFRP laminate always in contact with each other. For calibrating the frequency deviation meter's readings in term of fibre volume fraction, a number of CFRP laminates were made using compression moulding technique. Commercially available carbon fibre "pre-pregs" were used for making CFRP laminates. Pre-pregs consisted of type 1 (High modulus–surface treated, HM–S) carbon fibres in semi-cured DX-209 epoxy resin. Number of pre-pregs used for making the laminates were generally 10 for the technique used in this study. However, a few laminates were made using 8, 9, 11 and 12 layers of pre-pregs to get a wide variation in fibre volume fraction from laminate to laminate. All the laminates were scanned using the system described earlier and their eddy-scans were obtained. Areas having different meter readings were marked and sections were cut to have a number of samples having different meter readings. These sections were then mounted in cold setting epoxy resin for micrographic polishing. After polishing the samples, an image analysis system was used to find the area count of carbon fibres in different cut sections. The area count which was taken at minimum of 10 points on each sections, provided the fibre volume fraction readings and calibration curve was then obtained by plotting the average of obtained fibre volume fraction values against their respective meter readings (Fig. 3.9). This calibration curve supports the theoretical analysis which yielded that the variations in fibre volume fraction always affect the reactance of the probe and thereby frequency deviation meter's reading. This study has thus provided a non-destructive test method for measuring fibre volume fraction in carbon fibre reinforced plastic composites.

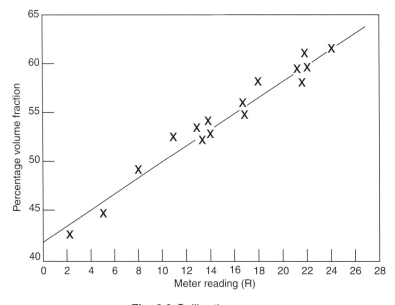

Fig. 3.9 Calibration curve

The calibration curve shown in Fig. 3.9 can be expressed mathematically by the following expression:

$$V_f = 0.8R + 42$$

where R is the reading of the frequency deviation meter used. It can be observed from this equation that for R = 0 (*i.e.*, when there is no deflection of frequency deviation meter's pointer), V_f = 42% and for R = 25 (*i.e.*, for full deflection of frequency deviation meter's pointer), V_f = 62%. Hence, the system described above is capable of measuring variations in fibre volume fraction in the range of 42% to 62% only. It may, however, be pointed out here that these are the practical limits of fibre volume fraction in CFRP laminates any way. CFRP laminates having V_f less than 42% are found to be weak due to presence of resin rich areas, which are potential failure sites and CFRP laminates having V_f higher than 62% are found to possess very poor transverse strength and transverse stiffness properties, due to stress magnification in thin resin rich areas.

Parameter V_f plays an important role in characterising mechanical behaviour of fibre composites. In fibre composites, it is fibres which bear the load and resin matrix just supports the fibres. In absence of correct information about fibre volume fraction, mechanical behaviour of fibre composites cannot be correctly predicted for designing structural components.

For calibration purposes, fibre volume fraction can be evaluated using "burn-off" technique or "wet-combustion process" also, if image analysis system is not available. The wet-combustion process and other methods such as acid digestion and quantitative microscopy are described in available literatures. For want of satisfactory non-destructive testing technique, "burn-off" technique is widely used for determination of fibre volume fraction. In "burn-off" technique, a small piece of laminate known as "cupon" is weighed and kept in a crucible. Thereafter, the cupon is burnt at a specific temperature by placing the crucible in a furnace. In the burning process, resin gets evaporated leaving carbon fibres in the crucible. These fibres are subsequently weighed to obtain fibre weight fraction, which is converted to fibre volume fraction using the following expression:

$$V_f = W_f \cdot \frac{\rho_c}{\rho_f}$$

where V_f = fibre volume fraction,

$\quad W_f$ = fibre weight fraction,

$\quad \rho_c$ = density of composite, and

$\quad \rho_f$ = density of fibre.

Value of ρ_f is generally provided by the manufacturer of fibre and ρ_c is determined experimentally using a density bottle or a density gradient column.

It is not uncommon to observe variation in fibre volume fraction from place to place on same laminate. This variation could be as high as 7 to 9%. In such a situation, evaluation of fibre volume fraction using a small cupon, cut from one of the edges of laminate, would provide a highly unrepresentative value of fibre volume fraction at some other point on the same laminate, say at a critical section. Therefore, it would not be wrong to state that destructive testing techniques (such as burn-off technique or wet-combustion process) for determination of fibre volume fraction in CFRP, are not only destructive and tedious but unrepresentative as well. The eddy-current method described above is non-destructive as well as it provides the information about fibre volume fraction (the exact reading) at the location of interest itself.

3.12 DETERMINATION OF LAY-UP ORDER IN CROSS-PLIED CFRP LAMINATES

As was mentioned earlier, a non-circular eddy-current probe will tend to give a non-circular eddy-current path even in isotropic conductors and the induced current path for such a probe varies with rotation of probe. These changes in induced current cause the impedance of probe to vary. The variation depends upon shape of probe and anisotropy of material under test. Consequently by using a standard shape probe, one may study variations in anisotropy. For CFRP, if one uses a non-circular probe, the impedance of the probe would be found to change with rotation angle in a manner determined by the nature and lay-up order in composite.

The reactance sensitive eddy-current circuit which was used for non-destructive evaluation of fibre volume fraction in unidirectional CFRP composites and which has been described in the previous section was slightly modified (Fig. 3.10) for orientation studies. In this circuit too, reactance of the probe is monitored. The probe has been made the inductance

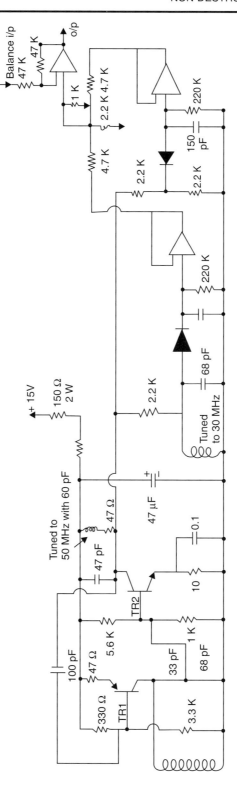

Fig. 3.10 Eddy-current circuit for orientation studies

of tank circuit of a free-running oscillator. Feedfack on TR1 maintains the oscillation, while TR2, an emitter follower gives an impedance match to the measuring circuit. A change in the reactance of the probe changes the frequency of oscillator and a frequency discriminator converts this frequency change into an equivalent direct current voltage.

To gain appropriate angular discrimination, a horse-shoe probe was constructed by gluing together several ferrite rings (3.2 mm nominal outside diameter) and to get an increased length of the probe, non-magnetic spacers were placed between consecutive ferrite rings. A length of 14 mm was thus obtained. The horse-shoe shape was obtained by grinding this tube on one side using fine wet emery cloth. The separation between probe faces thus obtained was 1 mm. Three turns of copper enamel wire were wound onto the probe.

The output of the frequency discriminator circuit was fed to a recorder, synchronised with the rotation of the probe so as to give a plot of output voltage versus the angle of relative rotation. Figures 3.11 and 3.12 show typical traces thus obtained. As the number of cross layers changes, the relative amplitude of horizontal and vertical wings changes.

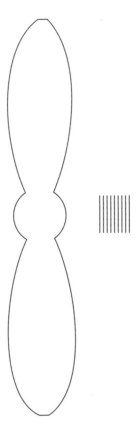

Fig. 3.11 Trace obtained with horse-shoe probe for unidirectional CFRP laminate

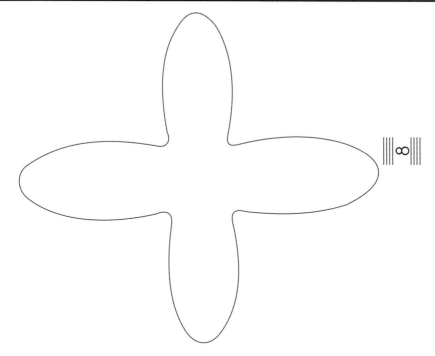

Fig. 3.12 Trace obtained with horse-shoe probe for a cross plied CFRP laminate

As mentioned in the previous section, the electrical analogue of eddy-current path in a layer of CFRP consists of a classical RLC (Resistance-Inductance-Capacitance) circuit.

A cross-plied laminate is made from a sequence of sheets of pre-pregs. With any pre-preg layer, the fibres in adjacent planes are parallel but at the interface between pre-preg layers of cross-plied orientation, the fibres in adjacent planes are perpendicular. At such an interface, this crossing of the fibres provides an easy resistive path for the induced current regardless of direction in the plane. Consequently the eddy-current circuit for such an interface reduces from an RLC circuit to an RL circuit. The series secondary impedance in these two cases is given by

$$Z_s = j\omega L_2 + R_2 + \frac{1}{j\omega c_2} = Z_{RLC} \qquad \qquad ...(21)$$

$$Z_s = j\omega L_2 + R_2 = Z_{RL} \qquad \qquad ...(22)$$

At first sight, a cross-plied laminate should be equivalent to several RLC circuits in parallel representing the layers of pre-preg, with several RL circuits in parallel representing the interfaces.

The presence of interface with fibres crossing can also be expected to influence the conduction processes for some distance on either side of the interface. The distance to which influence may extend is difficult to determine and one is forced to make assumptions which can either be accepted or discarded on the basis of the fit between experimental observations and theoretical predictions.

As a starting point it was assumed that if a pre-preg layer was bounded on both sides by layers whose fibre direction is parallel, the electrical behaviour would be characterised by R,L and C, *i.e.,* by Z_{RLC} equation (21), whereas if the layer is bounded on both sides by layers whose fibre direction is perpendicular, the behaviour would be characterised by only R and L, Z_{RL}, equation (22). For layer bounded on one side by a parallel layer and on the other side by a perpendicular layer, the situation is more complex and for the sake of simplicity it was assumed that such layers do not lose their RLC characteristic significantly and equation (21) still hold good. Interfaces will have their own electrical behaviour and their impedance will be designed by Z_1.

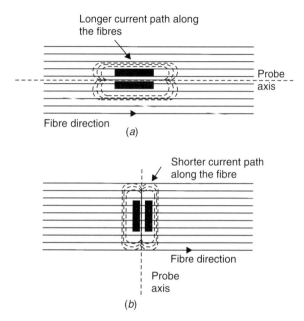

Fig. 3.13 Dependence of values for R, L and/or C on relative orientation
of the probe and fibre direction in layer concerned

For a particular probe, the values to be taken for R, L and/or C will depend also on the relative orientation of the probe and the fibre direction in the layer concerned (Fig. 3.13). This gives four cases, as shown in Table 1. These are in addition to the interface impedance Z_1.

The impedance of the primary (probe impedance) in the presence of the secondary (material) is given by

$$Z_p = j\omega L_1 + \frac{\omega^2 M^2}{Z_s} \qquad\qquad ...(23)$$

where M = mutual inductance.

Table 1. Values for impedance, for various combinations of
lay-up and arrangements of probe axis

	Pre-preg layer bounded by parallel layers on one or both sides	Pre-preg layer bounded by perpendicular layers on both sides
Probe axis parallel to fibres	Z_{RLC}	Z_{RL}
Probe axis perpendicular to fibres	Z'_{RLC}	Z'_{RL}

Using the above assumptions, the values of Z_s for blocks or materials made with various lay-ups for a particular position, *i.e.*, when the probe axis is perpendicular to the fibres of top layer are:

Case 1: Nine unidirectional layers or nine impedances in parallel.

$$\frac{1}{Z_s} = \frac{9}{Z'_{RLC}} \quad [11111111] \tag{...(24)}$$

Case 2: One 90° (cross) layer and eight 0° layers.

$$\frac{1}{Z_s} = \frac{8}{Z'_{RLC}} + \frac{1}{Z_{RL}} + \frac{2}{Z_1} \quad [111101111] \tag{...(25)}$$

Case 3: Two cross layers and eight 0° layers.

$$\frac{1}{Z_s} = \frac{8}{Z'_{RLC}} + \frac{2}{Z_{RL}} + \frac{2}{Z_1} \quad [1111001111] \tag{...(26)}$$

Case 4: Two cross layers and eight 0° layers.

$$\frac{1}{Z_s} = \frac{8}{Z'_{RLC}} + \frac{2}{Z_{RL}} + \frac{4}{Z_1} \quad [1110110111] \tag{...(27)}$$

Expressions for other lay-ups can similarly be written. For a position when the probe axis is parallel to the fibres of top layer, for case (1), equation takes the form:

$$\frac{1}{Z_s} = \frac{9}{Z_{RLC}} \tag{...(28)}$$

The change in output voltage (δV) from the discriminator is proportional to change in frequency.

At the working frequency of the apparatus one is above the natural resonant frequency of the induced eddy-current circuit in the unidirectional, *i.e.*, frequency is greater than that given by

$$\omega = \frac{1}{(LC)^{1/2}} \tag{...(29)}$$

Under these conditions the secondary (material) has an inductive reactance and the effect on the primary (probe) is to reduce the primary inductance (equation 23) which is a first approximation in the particular situation of eddy-current testing CFRP, has the form.

$$Z_p = j\omega L - KX_s \tag{...(30)}$$

where K is a constant and X_s represents the secondary reactance. X_s is related to Z_s but depends on the frequency as well as on L and C.

The changes in Z_p produced by the presence of the material are less than 10%. Although the running frequency of the oscillator in the test circuit is proportional to $1/(L_p)^{1/2}$, the change in the frequency for a small change in inductance is proportional to $1/\delta L_p$ and hence the output voltage from the equipment is proportional to $1/\delta L_p$. Putting the above argument together, it can be said

$$\delta V \propto \frac{1}{Z_s} \tag{...(31)}$$

where δV is the difference in output voltages for a specimen in position on the probe and no specimen present on the probe.

Using the equations for $1/Z_s$ for various cross-plied CFRP laminates (*e.g.*, equations 24–28) and the experimental voltages δV for these laminates it was possible to calculate values of Z_{RLC}, Z_{RL} etc. in relative terms (equation 31).

Using these experimental values for Z_{RLC}, it was possible to calculate expected voltage changes, δV, for any other laminate, theoretical and experimental values are shown in Fig. 3.14). When there were only two interfaces the fit between theoretical prediction and experimental result was very good but when there were more than two interfaces, there was no agreement. Further consideration showed that a fit between theory and experiment could be obtained if

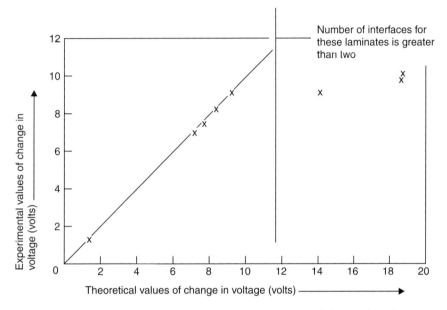

Fig. 3.14 Comparison of expected and actual values of change in voltage due to the presence of a cross-plied CRFP laminate

(*a*) The component $1/Z_s$ for two interfaces, *i.e.*, $2/Z_1$ (equation 25), was that initially determined, *i.e.*, 4.8 units.

(*b*) The additional effect of two more interfaces was 1 unit, making a total of 5.8 units (not the $2 \times 4.8 = 9.6$ units to be expected from the theory developed above).

(*c*) For a further two interfaces, *i.e.*, 6 in all, the additional effect was about 0.1, making 5.9 units in total $(4.8 + 1.0 + 0.1)$.

(*d*) For a further two interfaces, *i.e.*, 8 in all, the additional effect was again about 0.1 making 6.0 in all.

Figure 3.15 shows the experimental and theoretical values used in this calculation. The fit is very good.

Because this empirical modification was necessary, the theory developed above cannot be the whole story, even if it represents a part of it. Why the interfaces should be successively less effective is not clear. Two possible explanations are that once there are some conducting interfaces, the addition of others does not greatly affect the conduction pattern or that one interface, being a good conductor, screens the rest of the material from influencing the probe. This latter explanation seems to be unlikely because, with all the samples investigated, the probe could readily detect the presence of metal objects on the remote side of the laminate, *i.e.*, there was always complete penetration.

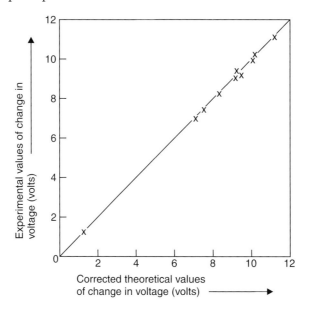

Fig. 3.15 Comparison of expected and actual values of V

3.13 CONCLUSION AND FUTURE POTENTIAL

In conclusion to lay-up order determination techniques, it may be said that it is possible to devise an eddy-current technique which can determine fibre lay-up in a CFRP structure. The system has been experimentally demonstrated and partially successful theoretical treatment has been offered. The system is worth further theoretical study to clarify the mode of operation, and this might allow more information regarding the structure, to be extracted from the experimental observations.

The technique can also be used to indicate the presence or absence of fibre wash in CFRP laminates. Where there is fibre wash, the characteristic loop for the materials is not obtained. The differences between the characteristic loop of a good laminate and the loop from a laminate containing wash may give a method of quantifying the degree of fibre wash, even when it is invisible, below the surface. Such information is important as fibre wash affects the strength and stiffness of structures.

CHAPTER **4**

Magnetic Particle Flaw Detection

For detection of surface and sub-surface flaws in ferromagnetic materials (such as common steels, iron, nickel, cobalt etc.), one should go for magnetic particle flaw detection technique. This technique is simple to use and provides rapid and unambiguous results. The sensitivity of this technique is maximum for surface flaws. For subsurface flaws (*i.e.*, flaws which are very near to the surface but are not open to the surface) also, this technique may be used. The sensitivity of this technique diminishes very rapidly as the depth of flaw below the surface increases. Hence, this method cannot be used for detection of internal flaws. Magnetic particle inspection technique cannot be used for non-ferromagnetic materials such as brass, copper, aluminium, titanium, magnesium, bronze, lead, ceramics, stainless steel etc. because these materials cannot be magnetised.

4.1 PRINCIPLE OF MAGNETIC FLAW DETECTION

Magnetic materials provide a preferential path for the flow of lines of magnetic flux. The flux flows from south pole to north pole within a magnet and flux flows from north pole to south pole, preferably through a magnetic material, outside the magnet, as shown in Fig. 4.1.

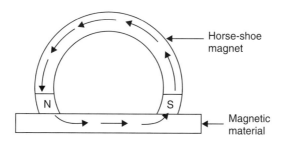

Fig. 4.1 Path of magnetic flux

Surface and sub-surface cracks in a ferromagnetic material produce distortions in the applied magnetic field. As the surface/sub-surface flaws are nothing but air gaps (non-magnetic), the lines of magnetic flux do not prefer to go through the air gap provided by these flaws and they try to follow the preferential path, *i.e.,* through the magnetic material under test. However, in the process of doing so, these lines of magnetic flux are, nevertheless, forced out of material. The "forcing out" is caused by the mutual repulsion between these magnetic lines of flux and those already present in the material (Fig. 4.2). The "forced-out" flux is commonly referred to as "leakage flux" and it is this "leakage flux" on which the principle of magnetic particle flaw detection is based. In all types of magnetic particle flaw detection technique, this "leakage-flux" is either measured or just detected.

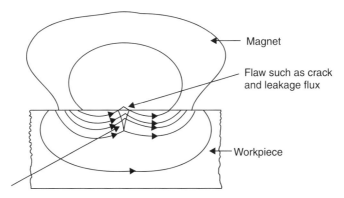

Magnet

Flaw such as crack and leakage flux

Workpiece

Fig. 4.2 Magnetic flux and leakage flux

It is a well observed phenomenon that when a bar magnet or a horse-shoe magnet or alike is taken near iron filings or other magnetic particles, the filings/particles cling to the poles of the magnet. This phenomenon is nothing but the simple fact that the magnetic particles are attracted by "leakage-flux" at the poles. Therefore, the best way of detecting "leakage-flux" is to take help of magnetic particles and that is why, this technique is called magnetic particle flaw detection technique.

There is another crude but simple way of explaining the principle of magnetic particle flaw detection. As is well known, if one breaks a bar magnet into two pieces, one ends up with two small size independent magnets (*i.e.*, at the broken edges, proper polarity develops, as shown in Fig. 4.3). Similarly, if a magnetic material having no surface/sub-surface flaw is magnetised; polarity shall develop only at the edges of the material under inspection. However, if a cracked piece is magnetised, polarities develop at the edges of material as well as at the crack (Fig. 4.3). Hence, when magnetic particles are spread over the magnetised workpiece; the particles get attracted not only at the edges but to the cracks as well. The magnetic particle "deposit" on the crack provides an unambiguous indication regarding the presence of crack.

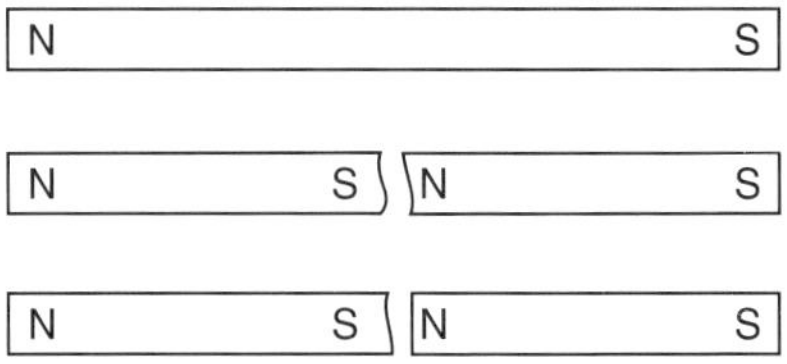

N	S

N	S	N	S

N	S	N	S

Fig. 4.3 Development of polarity on breakage

4.2 TYPES AND METHODS OF MAGNETISATION

There are two types of magnetisation, *viz.* circular magnetisation and longitudinal magnetisation; which are commonly employed for magnetizing a material for subsequent flaw detection. While selecting a particular type of magnetising, one should always remember that the best indication (say in the form of well-defined powder pattern) is obtained only when the magnetic lines of flux are at right angles to the crack. This is so because the cracks which are at right angles to the magnetic lines of flux, intercept more lines of flux and cause stronger leakage flux as compared to that provided by cracks which lie more or less parallel to the lines of flux.

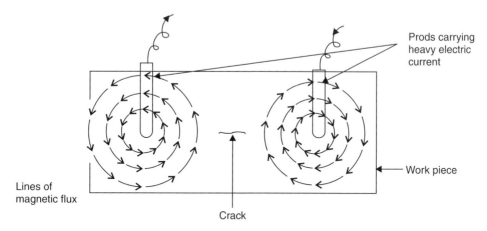

Lines of magnetic flux

Prods carrying heavy electric current

Work piece

Crack

(a) Local circular magnetisation

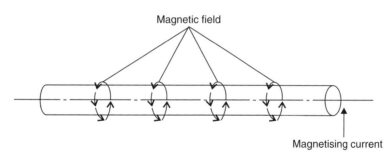

Magnetic field

Magnetising current

(b) General circular magnetisation

Fig. 4.4 Circular magnetisation

Regarding methods of magnetisation of a workpiece; most commonly used method for creating magnetic field in magnetic materials is to use high amperage electric current, as shown in Fig. 4.4 (a) and (b). While using electric current for magnetisation, one has got the advantage of changing the direction of lines of magnetic flux by suitably changing the direction of electric current. Also depending upon the requirement, one may place the workpiece in the field of a coil carrying heavy electric current (or solenoid) for obtaining induced magnetic field (Fig. 4.5). Electromagnets or permanent magnets (Fig. 4.6) or magnetising yokes (Fig. 4.7) may also be used for magnetisation of workpiece. Indian Standard IS: 7743-1975 recommends that A.C. electromagnetic yokes should be used for magnetisation only if the yoke has a lifting power of at least 4.5 kgf and pole spacing is 75 to 150 mm. D.C. electromagnets or permanent magnetic yokes are recommended for magnetisation only if such yokes have a lifting power of at least 18 kgf and pole spacing of 75 to 150 mm.

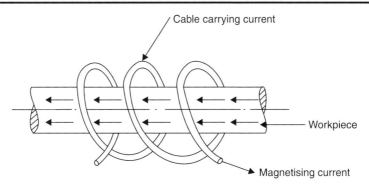

Fig. 4.5 Longitudinal magnetisation

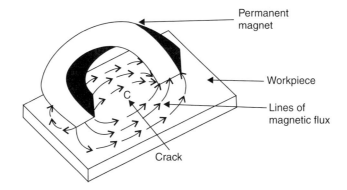

Fig. 4.6 Magnetisation by permanent magnet

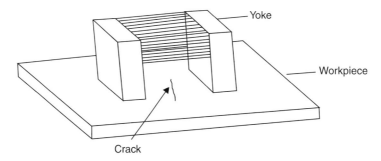

Fig. 4.7 Magnetisation by yoke

Magnetisation which is shown in Fig. 4.4 (*a*) and (*b*) is circular magnetisation and those shown in Figs. 4.5, 4.6 and 4.7 are longitudinal magnetisation. For obtaining the direction of magnetic lines of flux, right hand rule applies. According to this rule, if one points out his right hand's thumb in the direction of magnetising current; then the folded fingers show the direction of magnetic lines of flux (Fig. 4.4). If the current carrying material is magnetic too; the magnetic field is induced in the material, as well as in the surrounding space. However, in case of non-magnetic materials, lines of magnetic flux are induced only around the current carrying conductor.

If a high amperage current is passed through a coil of several turns; longitudinal magnetic field (Fig. 4.5) is established within the coil. The nature and direction of this field are the result of the field around the conductor which forms the turns of the coil. If one applies the right hand rule to the conductor (at any point on the coil) carrying electric current, it will be found out that a longitudinal field is established within the coil.

Whether, the component to be inspected should have circular magnetisation or longitudinal magnetisation, depends primarily on the direction of likely cracks. As mentioned earlier, for reliable inspection, the lines of magnetic flux should be at right angles to the crack to be detected. Hence, circular magnetisation is required for detecting cracks which are parallel to the length of workpiece and longitudinal magnetisation is required for transverse cracks.

4.3 MAGNETIC PARTICLES

Magnetic particles are not just iron filings but are made from selected magnetic materials (such as finely divided iron oxide powder) of proper size, shape, magnetic permeability and retentivity. Sometimes, these particles are coloured to provide a contrast between the particles to be observed and the surface to be inspected. Generally, these particles are either black or grey but sometimes red coloured magnetic particles are also used for obtaining improved contrast.

In certain applications, instead of using coloured magnetic particles, the surface of the workpiece to be inspected is painted with quick-drying white paint for enhancing the contrast with usual black or grey coloured magnetic particles. It may be mentioned here that as these paints are poor electrical conductors; it becomes very difficult to pass electric current through such paints to the metallic surface and hence, either the paint is partially removed from areas where magnetising prods are to make electrical contact, or one is required to go for coil magnetisation. For hollow parts, use of a central conductor may be made for magnetisation without removing the white paint. Sometimes, to improve the conductivity of high amperage current through paints; titanium or magnesium based paints are used and only a very thin coating is applied to the surface.

4.4 DRY AND WET METHODS OF MAGNETIC PARTICLE INSPECTION

Magnetic particles are applied to the magnetised surface under inspection, either using air as a carrying agent or using proper liquid base for transportation of the particles to the surface. If the particles depend upon air to carry them to the surface, the technique is called "dry method" and if a liquid base (say thin transparent paraffin) is used, the method is called "wet-method".

In dry method, dry magnetic particles which are either natural colour (*i.e.,* grey or black) or which have been colour/fluorescent coated are used. Magnetic properties and size of magnetic particles are similar irrespective of whether they are grey, black, red or fluorescent coated. Hence, it is the criterion of "best-contrast" which dictates the choice of a particular magnetic powder. For *fluorescent coated magnetic particles*, viewing is done under ultraviolet/black light source.

While applying magnetic powder to the surface, one must remember that the powder should not be just dumped over it. When the powder gently floats to the surface, the magnetic particles are freely influenced by the "leakage flux" to form correct indication of the crack; whereas in case of dumping of magnetic powder on the surface, they are not freely influenced by the leakage flux, resulting in incorrect indications.

Dry magnetic powder can be applied using commercially available rubber spray bulbs or mechanical powder blower. Another simple way out is to let finely divided magnetic powder fall over the surface freely and it can be achieved by making the magnetic powder sieve through a fine cotton cloth.

In wet method, as mentioned earlier, fine magnetic particles are suspended in transparent paraffin based liquid. This liquid is sometimes referred to as magnetic ink. Wet particles are best suited for detection of fine surface cracks such as fatigue cracks. They are generally used in stationary equipments where a bath containing magnetic fluid may be used. Generally, magnetic paste (available in black, red or fluorescent particle coatings) is used for dissolving in suitable oil/liquid base to prepare the magnetic bath. The bath thus prepared contains suspended magnetic particles in liquid base.

As was mentioned for dry magnetic powders; the choice of a particular magnetic paste too depends entirely upon the criterion of maximum contrast. Also, indications formed by black or red pastes may be viewed under ordinary light and fluorescent paste indications are viewed under black light/ultraviolet light source.

Magnetic pastes may be dissolved in water too to obtain suspension of magnetic particles. However, as water may corrode the magnetizing equipment/workpiece and also because water does not wet the workpiece thoroughly, use of water as liquid base is very restricted.

4.5 USE OF FLUORESCENT COATED MAGNETIC PARTICLES

Fluorescent coated magnetic particles are widely used for magnetic particle inspection. When fluorescent coated magnetic particles are used, their presence is not brilliantly indicated when viewed under ultraviolet/black light source. Commercial black light sources produce black light (near ultraviolet light) of a specific wavelength (3650 Å). This wavelength is in between the visible and ultraviolet range and is considered as non-injurious to skin and eyes. Generally, the blacklight source consists of a 100 water reflector-spot-type of mercury vapour bulb alongwith a transformer and a filter. The filter absorbs visible light and allows only the light of proper wavelength (i.e., only the so-called safe-light). Fluorescent particles fluoresce under black light (i.e., the magnetic particles adhering to the crack, appear as a shining yellowish-green line).

The main advantage of fluorescent particles is their increased visibility under black light. Fluorescent indications are one hundred times more easily visible than black or grey or red particles indications. For mass inspection, fluorescent indications permit inspection speeds as much as six to ten times than those possible with other magnetic particles.

4.6 INDUSTRIAL APPLICATIONS

Magnetic particle flaw detection technique is widely used in various industries at different stages, such as for receiving inspection, for inprocess inspection, for final inspection etc. Generally, it is employed at all the three aforementioned stages. The technique is used by the Quality Control and Inspection Department for varied applications and it includes routine inspection and preventive maintenance of plants and machineries.

The technique is widely used for inspecting crank-shafts, connecting rods, pipes, tubes, castings, weldments, springs, threaded parts, forgings etc. It is successfully used for detecting non-metallic inclusions, cracks filled with foreign materials, stringers lying close to surface, presence of seams, thermal and heat-treatment cracks, cooling and quenching cracks, forging laps, forging burst and flakes, laminations, incomplete weld penetrations, sub-surface blow-holes, grinding and machining cracks, fatigue cracks etc.

In brief, magnetic particles flaw detection technique is a very sensitive technique and a rapid means of locating small and shallow surface cracks in ferromagnetic materials. There is no limitation to the size and shape of the part being tested. Also, generally, there does exist the requirement of surface preparation or elaborate cleaning of the surface for carrying out magnetic particle inspection. Discontinuities at subsurface levels can also be located by this method, though only to a limited extent. For sub-surface discontinuities, dry method should be used for obtaining better indications of the discontinuities. Deeper discontinuities do not produce a readable indication and, therefore, cannot be identified.

4.7 WORKING OF A FEW COMMERCIALLY AVAILABLE MAGNETIC CRACK DETECTORS

A. Prod Type Magnetic Crack Detector

Most of the portable type of commercially available magnetic crack detectors are of prod type. Prod type magnetic crack detectors basically have a transformer unit to provide high amperage electric current, two electric prods for carrying electric current from transformer unit to the component to be magnetised and back to the transformer unit, sufficient cable lengths for making coil if required, holding arrangement for the prods and a conveniently situated "push-button" type electric switch for switching "on" and "off" the flow of electric current through the component under inspection. This "push-button" switch is generally hand switch situated near the operator's right hand thumb. It could also be a pedal switch situated near operator's foot. Whether it should be a hand switch or a pedal switch, depends upon whether an operator can easily switch-on or switch-off the unit within seconds.

In prod type magnetic crack detectors, it is absolutely necessary to hold the prods in an upright position and to hold it tight on the component's surface. If the prods are not held upright and tight; air-gap may exist between the prod and the surface of component. Due to this air gap, arcing and sparking may take place. (It is because of high amperage electric current flowing from the prod to the component). Due to arcing, the component's surface may get spoilt at the point of contact and the operator may, due to sparking, get burnt-holes in his clothes. He may also injure himself. Hence, improper contact, resulting in arcing, should be avoided by all means. Shaky hands also result in sparking and therefore the operator should hold the prods with steady hands. If the points where prods are to make contact are not properly cleaned, that too results in arcing because the dirt particles do not let the prod sit properly over the component's surface, which results in arcing. Therefore, the area where prod is to make contact with the component's surface, must be cleaned properly and the prod should be held tightly and upright over the surface under inspection. Electric current is switched-on only when the prods have been firmly placed over the surface.

Using the prods, high amperage electric current is passed through the component, thereby magnetising the component. Magnetic powder is then blown over the surface. For best results the powder should be blown over the surface while magnetising current is still "on". However, reasonably good results are obtained even when the powder is sprayed over the surface after switching "off" the magnetising current. In the later case, formation of powder pattern, on the surface of component under test, is due to residual magnetism. Powder pattern due to residual magnetism is usually not as sharp and as well-defined as is in the case of spraying of magnetic powder while magnetising current is still on.

After magnetic powder has been sprayed over the entire magnetised surface with the help of a rubber spray bulb or alike; the magnetising current is switched off and the prods are removed from the surface of the component. Thereafter, the component is turned upside down and is given a gentle tap. Magnetic particles lying loose on the surface fall-off due to gravity leaving only those magnetic particles on the surface which are being held tightly on the surface by the leakage flux due to presence of crack etc. The magnetic-particles deposit or the so-called "powder pattern" on the surface shows the presence and location of the cracks, unambiguously. This powder-pattern, however, does not provide any quantitative information regarding the size of the crack or the depth of the crack. An experienced operator may nevertheless provide an approximate idea regarding the size of the crack.

Prod type magnetic crack detector units vary from small low current units to medium and heavy duty equipment. Whereas small units supply magnetising current of the order of 500 amp.; higher current units supply 1,000 to 10,000 amp of A.C., D.C. or half-wave rectified magnetising current. For inspection of medium sized weldments and castings, the equipment most widely used supplies A.C. or half-wave current at the option of the operator. Sensitivity may be controlled to reveal only surface cracks with A.C. and both surface and sub-surface cracks with half-wave current.

All prod type magnetic crack detectors rated above 500 amp. operate from either 220 volt single phase or 440 volt three phase 50 c.p.s. A.C. lines. For outdoor applications, where there are no electrical connections, diesel generators may be used to tap 220 volts or 440 volts A.C. for the magnetic crack detector unit.

As shown in Fig. 4.4, when high amperage electric current is passed through prod to component, one obtains a local circular magnetisation. Maximum lines of magnetic flux of such magnetisation are intercepted by cracks which are along the line joining the prod contact points (Fig. 4.4) and cracks so oriented are easily detected. A crack which is across the line joining the prod contact points either does not intercept any line of magnetic flux or intercepts only a few lines of magnetic flux and, therefore, provides only a very feeble leakage flux. Hence, a crack which is perpendicular to the line joining prod contact points, evades its detection. Therefore, when using a prod type magnetic crack detector for locating surface cracks of unknown orientation, the component under inspection should be magnetised in two mutually perpendicular directions to avoid missing any unfavourably oriented crack. If the cracks to be detected are at angles other than at right angles to the line joining the prod contact points, they intercept sufficient number of lines of magnetic flux and, therefore, provide enough leakage flux for the magnetic particles to adhere to them and thus provide indication of the cracks present in the component.

For obtaining longitudinal magnetisation from prod type magnetic crack detectors, cable of one of the prod is wound into several turns to obtain a coil and then the two prods are put in contact with each other firmly. This completes the path for the flow of high amperage current. The component under inspection is then placed in the field of coil to get longitudinal magnetisation of the component (Fig. 4.5). While the magnetising current is still flowing in the cable-coil, magnetic powder is sprayed over the surface of magnetised component. The component is then gently tapped to remove the loosely lying particles from the surface and only those magnetic particles, which are held to the surface by the leakage flux due to presence of cracks, remain on the surface. This provides the required indication of the presence of cracks at different locations.

Instead of spraying usual dry black or grey coloured magnetic powder, one may go for fluorescent coated magnetic particles (Section 4.5) or may adopt wet method (Section 4.4). In either case, the rest of the test procedure such as magnetisation etc. remain the same.

B. Yoke Type Magnetic Crack Detector

As already shown in Fig. 4.7, electromagnetic yokes may be used for magnetisation of components. Yoke magnetisation is described in section 4.2. Electromagnetic yokes provide a high intensity unidirectional magnetising field between the poles, when it is operated from 220 volts, 50 c.p.s. A.C. line or it may be used as a D.C. yoke when operated from a 12 volt car battery drawing 12 amp current. Magnetic powder is sprayed over the magnetised surface while the yoke is still getting energised to obtain good indications. As shown in Fig. 4.7, a crack which is at right angles to the line joining the poles of yoke, intercepts maximum number of lines of magnetic flux resulting in a strong leakage flux. Hence, if the expected orientation of the crack is known, then the yoke should be kept on the component in such a way that the line joining the poles (or the pole axis) is at right angles to the expected crack direction. In case of unknown crack orientations, inspection should be carried out in two stages with pole axis in two mutually perpendicular directions. By doing so, all possible crack orientations are taken care of, as has already been described for prod type magnetic crack detector.

C. Stationary Magnetic Particle Inspection Units

Stationary wet type horizontal magnetic particle inspection units are widely used for the inspection of small manufactured parts. These units are generally provided with a built-in-tank for holding magnetic fluid and pumping magnetic fluid over components, surface through hand-held hoses and for agitating the magnetic fluid bath. The component to be inspected is clamped between copper contact faces of the head and tail stocks within the magnetising coil. One may magnetise the component either longitudinally by setting the current pass through the coil or one may obtain circular magnetisation by letting the current pass between the heads. If required, one may get circular and longitudinal magnetisations simultaneously. While the part is still being magnetised, magnetic fluid is sprayed over the component's surface and indications thus obtained are viewed. Generally, magnetic fluid used with these stationary units is fluorescent magnetic ink and, therefore, viewing of indications is done under black light (*i.e.*, safe ultraviolet light). As one needs a dark area for viewing fluorescent indications, generally these units are provided with inspection hoods, which can be lowered to obtain a dark inspection area. As described earlier, by viewing fluorescent indications under black light in a dark area, the speed of inspection increases and chances of missing an indication are reduced to a minimum.

These stationary magnetic particle inspection units are operated from 220 volts A.C. 50 c.p.s. electric supply and these units vary in size and current ratings depending upon requirement. The size of unit decides as to how big a component can be accommodated in the unit. Some of these units can accommodate components of size up to 1 metre and the larger units can accommodate components size up to 2.5 metres. Regarding current ratings, these units have current rating in the range of 500 amp to 3,000 amp for magnetisation of the components under inspection. The amperage of magnetising current is controlled by a multiple point tap switch. Since these stationary magnetic particle inspection units are A.C. units, the components under inspection may be demagnetised, if required, using the magnetising coil of the unit itself and by decreasing the amperage of magnetising current in several steps.

Some of the stationary magnetic particle inspection units are D.C. type too. These units supply D.C. magnetising current. They are useful when sub-surface cracks too need detection besides the surface cracks detection. These D.C. units are provided with multiple-point automatic reversing D.C. tap switches for demagnetisation of components in place on the unit itself after the inspection.

D. Mechanised Inspection Units

Some of the magnetic particle inspection units are mechanised units and they are either semi-automatic or automatic in their functioning. The mass produced components are carried by continuous conveyer belts. Loading and unloading of these components are done either automatically or manually. The inspector inspects the components when these components pass through the test area and components having clear visible indications are rejected. Other components are deemed to have been accepted and they remain on the conveyer belt, which takes these components through an automatic demagnetiser unit for demagnetising these components. Thereafter, the components leave the inspection department. Such mechanised units permit rapid and low-cost inspection of mass produced components such as bolts, screws, power screws, connecting rods, cam shafts, gears, cams, similar automobile and mechanical components.

4.8 FLAW DETECTION IN RODS AND PIPES

Rods and pipes may have longitudinal cracks (*i.e.*, cracks along the length of rods or pipes), transverse cracks (*i.e.*, cracks, which are along the circumference of rods or pipes), both type of cracks or cracks having general orientation. Depending upon the likelihood orientation of the crack, one has to choose between circular and longitudinal magnetisation. Circular magnetisation is required for locating longitudinal cracks and longitudinal magnetisation is required for locating transverse cracks in rods and pipes.

For circular magnetisation of solid rods, the rod under inspection is placed between contact plates of head and tail stocks of a stationary type of magnetic particle inspection unit (Fig. 4.8). Thereafter, a high amperage electric current is passed through the rod, resulting in circular magnetisation of the rod. If there exists a longitudinal crack, it intercepts many lines of circular magnetic field and thus leakage flux is obtained which, in turn, provides the required magnetic particle indication of such cracks.

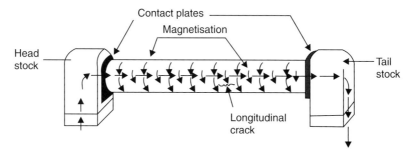

Fig. 4.8 Circular magnetisation of solid rods by passing high amperage electric current through it

For circular magnetisation of hollow rods, pipes or any other cylindrical part through which a central conductor (a copper bar or alike) could pass; one should take help of induced magnetic field which is obtained by sliding the hollow component over a central conductor, as shown in Fig. 4.9. The central conductor is clamped between the contact plates of head and tail stocks. High amperage magnetic current is then passed through the central conductor. This causes a magnetic field around the central conductor. Due to this, a magnetic field of sufficient intensity is induced in the hollow part too which has been slid over the central conductor. Passage of high amperage electric current in this fashion, induces circular magnetisation on both inside and outside the cylindrical surfaces of the hollow component. This way, one may inspect the inside surface too besides outside surface, for likely surface cracks etc. Instead of using the central conductor approach, if one goes for the magnetisation of such hollow components by directly passing electric current through them; one does not get satisfactory magnetic field on the inside surface and thus the inspection of inside surface has to be excluded. Therefore, wherever possible, hollow rods or pipes should be magnetised using central conductor technique (Fig. 4.9). This type of magnetisation which is also called threading bar method, enables one to detect the cracks which are parallel to the length of hollow rod or pipe, because such longitudinal cracks nicely intercept the circular magnetic field and provide leakage flux for the detection of such cracks by magnetic particles. If the magnetisation is accomplished by wrapping a few turns of cable to form a coil, around the hollow pipe, it is called threading coil method.

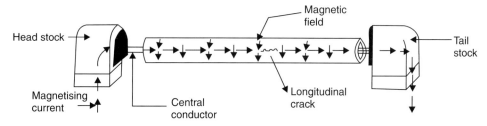

Fig. 4.9 Circular magnetisation of hollow rods and pipes by passing
high amperage electric current through a central conductor

Circular magnetisation with the help of a central conductor has the following advantages over the technique of directly passing the electric current through the part itself:

1. As the magnetic field is induced on the inside surface too, it becomes possible to inspect both the inner and outer surfaces of the hollow component for any probable defect.

2. Several parts such as washers, nuts, bushes etc. can be suspended on the central conductor and inspection can be carried out in groups.

3. As one does not go for direct electrical contact of the contact plates and the compo-nents, the likelihood of getting burnt surface of the components under inspection, is avoided.

Finally, a word of caution regarding incorrect practice of magnetisation by placing the component near the conductor rather than around it. As one obtains induced circular magnetisation by sliding a hollow rod or pipe over a central conductor carrying high ampere current; one may wrongly assume that a circular field may also be induced in a component, even by just placing it near the conductor carrying high amperage electric current. However, such a practice cannot be recommended because in doing so, a major portion of the magnetic field remains in air and only a small portion of it passes through the component under inspection, resulting in magnetisation, which is distorted, unevenly distributed and only partial. This type of "not-recommended" magnetisation is called parallel magnetisation and should never be practised.

For longitudinal magnetisation of rods and pipes, the rod or pipe under inspection, may be placed in the field of a coil carrying high amperage electric current (Fig. 4.10). As described earlier, when a high amperage electric current is passed through a coil of several turns, a longitudinal field is established within the coil (Fig. 4.5). The nature and direction of such magnetic fields are determined by the field around the conductor formed by the turns of the coil. If one applies the right hand rule to the conductor of coil at any point on the coil (Fig. 4.5), it shall easily be established that the field within the coil is lengthwise (*i.e.*, the component within the coil shall have longitudinal magnetisation).

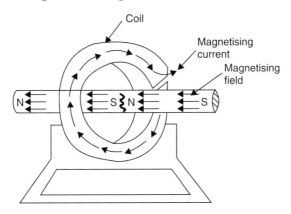

Fig. 4.10 Longitudinal magnetisation of rods and pipes by placing them
in the field of a coil carrying high amperage electric current

Longitudinal magnetisation of rods and pipes is carried out to detect transverse crack and transverse discontinuities. Such transverse defects nicely intercept the longitudinal magnetic field and the leakage flux is provided due to this interception, which provides the required indication by way of magnetic particle deposition on such defects. In other words, magnetic poles are formed on the two faces of the transverse cracks (Fig. 4.10), attracting magnetic particles to be held there and thus provide the required indication.

For longitudinal magnetisation of rods and pipes, one can even use the cables of a portable prod type magnetic crack detector. The cable of one of the prods is wound several turns to provide a coil and the rod/pipe is placed in the field of this coil. The two prods are then put in firm contact with each other, and high amperage electric current is allowed to pass through the cable, resulting in longitudinal magnetisation of the rod/pipe in the field of coil. This type of magnetisation is called cable magnetisation.

4.9 FLAW DETECTION IN A SHORT WORKPIECE

Flaw detection in short workpieces poses problem when using prod type magnetic crack detector. There is limitation regarding placing the two prods very close to each other because they carry very high amperage electric current and if they are placed too close to each other, it would result in arcing/sparking etc. Now, if the workpiece itself is short, the prods can naturally not be placed on the workpiece itself. Hence, for magnetisation and subsequent inspection of short workpieces, one needs to have a base material too which should be electrically conducting. The prods are placed at proper distance on the base material and the high amperage electric current passes through the prod to the base material and therefrom to workpiece. This way the workpiece comes within the local circular magnetisation due to prods and a crack, which is along the line joining the two prods, is easily detected. For other crack orientations, the workpiece should be rotated through 90° and the inspection carried out once again [Fig. 4.11 (*a*) and Fig. 4.11 (*b*)].

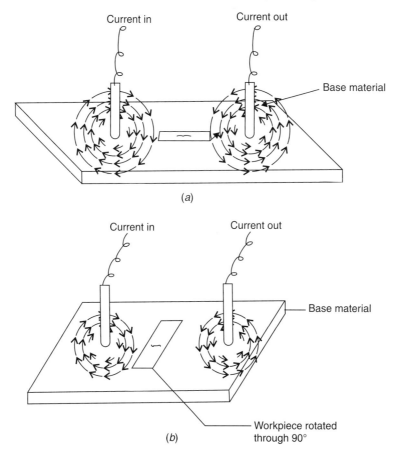

Fig. 4.11 Magnetisation of a short workpiece using prod type magnetic crack detector

For short workpieces having holes (such as nuts, washers etc.), one can go for their magnetisation by sliding them over a central conductor and passing a high amperage electric current through the conductor, as already described, while describing circular magnetisation of hollow rods and pipes.

4.10 PRECAUTIONS

(*i*) When employing prod type magnetic crack detector, the prods should be held steady and in upright position to avoid sparking. Sparking not only spoils the surface under inspection but may also cause burnt holes in operator's cloth and may cause minor injury to the operator. The places on the surface under inspection, where prods are to be kept, should be cleaned properly because presence of dirt etc. does not allow the prods to sit properly on the surface, resulting in sparking.

(*ii*) Depending upon the likelihood orientation of the crack, proper selection between circular and longitudinal magnetisation should be made.

(*iii*) When employing prod type magnetic crack detector for location of a crack whose likely orientation is not known; the inspection should be carried out in two mutually perpendicular directions either by rotating the workpiece by 90° for second inspection or by placing the prods along a perpendicular axis for the second inspection.

(*iv*) Magnetic particles should fall freely on the magnetised surface for the build-up of correct indications. The powder should not be just dumped over the magnetised surface.

(*v*) While using fluorescent particles, safe blacklight source should be used. Ordinary ultraviolet lights, which are injurious to skin and eyes, should not be used.

(*vi*) Fluorescent particle inspection should be done in dark area. The darker the area of inspection, brighter is the indication. Before, an inspector carries on the inspection in the dark area, he should be allowed adoption time to get used to dark area. (*i.e.,* about 5 to 10 minutes should be allowed after the inspector has come from bright or day-light areas, before the start of inspection).

(*vii*) If the component under inspection is short, a base material should be used for magnetization of the component. Prod spacing of less than 50 mm is not recommended.

(*viii*) If the component under inspection has to undergo subsequent machining or arc welding, or is to be used near instruments such as magnetic compass etc., the component must be demagnetised after the magnetic particle inspection is over.

(*ix*) Parallel magnetisation should not be practised.

(*x*) Non-relevant indications should be excluded and the indications must be properly interpreted.

4.11 LIMITATIONS

(*i*) The magnetic particle flaw detection technique is applicable only to ferromagnetic materials. Non-ferromagnetic materials such as brass, copper, aluminium, titanium, magnesium, bronze, lead, ceramics, polymers, fibre reinforced plastic composites, stainless steel etc. cannot be inspected using this technique.

(*ii*) The technique is capable of detecting only surface and sub-surface cracks. It cannot detect internal defects. Also, using A.C. type magnetisers, only surface cracks can be detected. For sub-surface cracks, one needs D.C. type magnetising units.

(*iii*) Generally, components need to be demagnetised after inspection. In such cases, further instrumentation is needed to check whether demagnetisation has been up to the requirement or not.

(*iv*) Sometimes, non-relevant indications are also present on the surface under inspection. Hence, if the indications are not properly interpreted; it may lead to rejection of good components. For example, at the junction of two different metals (*i.e.,* two metals having different magnetic permeability); magnetic particle indications are invariably obtained. This indication should not be interpreted as a defect.

(v) Magnetic particle flaw detection technique does not provide the exact size and severity of the flaws present. It only provides an estimate of the size and severity of the defect.

4.12 RESIDUAL MAGNETISM

Some of the components retain an appreciable magnetic field after inspection. This is known as residual magnetism. This does not affect the mechanical properties of the component and in many cases this will not be detrimental to subsequent usage. In some cases, however, this residual magnetism is undesirable and one is required to demagnetise the component to get rid of this residual magnetism. The need for demagnetisation and the usual techniques for demagnetising a component are given in the next section, *i.e.,* 4.13.

As described earlier, whenever possible, magnetic particles should be applied while magnetisation continues. This type of magnetic particle inspection is sometime called "continuous method". This continuous method provides better indications as compared to the method in which the magnetic particles are sprayed after the component has been magnetised and magnetisation current is no longer flowing. This type of magnetic particle inspection is sometimes called "residual method". The residual method is based upon residual magnetism and the effectiveness of residual method depends upon the strength of magnetising force and also on the magnetic characteristics of the material of the component. Residual magnetism also depends upon the geometry of the component and direction of magnetisation (*i.e.,* whether it is longitudinal magnetisation or circular magnetisation). Residual method should only be used in those cases where continuous method cannot be employed. Also, D.C. current should be preferred over A.C. current for residual magnetism. Sometimes, in continuous method, leakage flux is present, when the magnetising current is flowing, due to conditions other than defects. This is specially true when a current carrying conductor is wrapped around the specimen to be magnetised. However, in residual method, because of its lower sensitivity, the possibility of these false indications is eliminated.

4.13 NEED FOR DEMAGNETISATION

Careful consideration should be given to decide, whether or not, the component should be demagnetised, after magnetic particle inspection is over. In the following paragraphs, a few instances have been mentioned itemwise where demagnetisation is required. In industry, demagnetisation means reducing the degree of magnetisation to an acceptable level, because complete demagnetisation is usually impractical.

(i) If the component is to be subsequently machined, a residual magnetic field may cause chips to collect on the tool and adversely affect the cutting action, as well as, it would affect the surface finish. Hence, the components, which are to be subsequently machined, should undergo demagnetisation operation.

(ii) If the component is to be used in locations near sensitive instrumentations, high residual magnetic field may affect the operation of instruments. Even slightly magnetised aircraft parts may cause the magnetic compass of an aeroplane to read erroneously. Hence, in such cases, demagnetisation is a must.

(*iii*) When the part under inspection is to undergo subsequent arc welding operation, the presence of strong residual magnetic field is undesirable because this may deflect the arc.

(*iv*) Strong residual magnetic fields are source of excessive friction between moving parts; *e.g.*, between a piston and cylinder wall. Hence, proper precaution should be taken.

(*v*) If a rotating part has undergone magnetic particle inspection and has not been demagnetised; it may attract iron-filings, particles, chips etc. depending upon the environment in which it is rotating. This may result in malfunctioning; specially so in the case of bearing surfaces.

(*vi*) Residual magnetic field may also cause some problems, while cleaning a component which has earlier been subjected to magnetic particle inspection and has not been demagnetised. Due to residual field, chips, iron-filings, particles etc. keep adhering to the surface while spray cleaning is going on.

Some of the usual techniques which are commonly employed for the demagnetisation of components, after magnetic particle inspection, are described below:

(*a*) The method which is most widely used for demagnetisation of components is to withdraw the component very slowly from the field of high-intensity A.C. coil. A field strength of 5,000 to 10,000 Ampere-turns is recommended. Sometimes, instead of withdrawing the component, the coil is withdrawn slowly, keeping the part stationary. Care should be taken to remove the part entirely from the influence of the coil before the coil is de-energised. Otherwise, instead of demagnetisation, one will end-up with magnetisation of the part. This method is advantageous for high production rates, since a properly designed coil can be energised continuously while a steady stream of parts is conveyed through the coil.

(*b*) In another method, the alternating-current magnetising force is reduced, in steps, down to a negligible value. The reduction in field intensity is obtained by reducing the current to the coil, while the part remains within the coil, until the current is reduced to zero. Current control is achieved by various means, such as an autotransformer in conjunction with a tap switch, etc.

(*c*) The third method, which is known as reversing D.C. demagnetisation, consists of consecutive steps of reversed and reduced direct-current magnetisation down to a negligible value. This is the most effective method of demagnetising large parts. The A.C. field demagnetising process does not penetrate beyond the surface to remove the residual magnetisation and, therefore, this D.C. method which provides deep penetration is preferred over A.C. method. However, this method requires special equipment for reversing the current and simultaneously reducing it in 30 or so small decrements. If a coil is used, the part is left in the coil until the demagnetising cycle has been completed.

(*d*) The fourth method of demagnetisation is known as A.C. circular field demagnetisation. In this method, the demagnetisation current is passed through the component itself, instead of coil and the magnitude of the current is systematically reduced to zero using some suitable device. The method is useful for large parts just after their magnetic particle inspection. Some of the inspection units (stationary types) have built-in devices for the systematic reduction of current and usually bulky parts are demagnetised on the unit itself before their final removal.

(*e*) Yet another way of demagnetisation is to have A.C. and D.C. yokes. These yokes are suitable for the demagnetisation of small parts having high coercive forces. (After magnetisation of ferromagnetic materials, magnetic domains are oriented in a particular direction. The force which, when applied in opposite direction, causes oriented domains to return to their random orientation is called coercive force). These yokes are C-shaped and are designed for particular application. The component to be demagnetised is passed between the pole faces and then withdrawn slowly, till the component is completely out of magnetic field of the yoke.

It is easier to demagnetise a part which has been longitudinally magnetised, as compared to a part which has been circularly magnetised. After demagnetisation of the part, one should check whether or not the demagnetisation has been accomplished to the accepted level. This can be checked using field indicator, compass indicator, steel wire indicator etc. Sometimes, similar parts which have been scrapped, are used to check the effectiveness of demagnetisation process. For example, to evaluate the effectiveness of demagnetisation of a bearing race magnetised with a central conductor, one can do so by cutting the scrapped sample into two pieces. The presence of residual magnetism can then be checked by applying magnetic particles to it which will adhere to all the four exposed ends, if demagnetisation is not correctly achieved.

4.14 RELEVANT AND NON-RELEVANT INDICATIONS

As is obvious from the title itself; the indications which are due to defects, are classified as relevant indications and indications which do not pin point to a defect but all the same are present on the surface, are classified as non-relevant indications.

Surface discontinuities and sub-surface discontinuities such as quenching cracks, seams, undesirable non-metallic inclusions, laminations, forging laps, sub-surface blow holes or gas porosities, heat treatment cracks, grinding cracks, plating and etching cracks, fatigue cracks etc. all provide relevant indications. Out of these cracks, the ones which are "close-lipped" and sharp are easiest to observe because they end-up in a well-defined particle pattern. For sub-surface cracks, the powder pattern becomes broader and fuzzier as the depth of sub-surface crack below the surface increases. For the detection of sub-surface cracks, dry method has an edge over wet-method.

One of the sub-surface discontinuity, which is often located by magnetic particle inspection technique, is the non-metallic inclusions. Even though it is found most frequently, is often of little consequence as far as usefulness of the part is concerned. Only when such inclusions occur in areas of high stress, they may warrant the rejection of part. Therefore, it shall be unfair to include all the non-metallic inclusions in the category of relevant indications. Unless they are responsible for strength reduction, indications from them should be categorised as non-relevant.

There are certain other indications which are non-relevant irrespective of their position and loading condition. These indications cannot be categorised as "false indications" because they are caused due to leakage flux itself. However, such leakage fluxes are caused by factors which do not have any bearing on the suitability of the part. Such non-relevant indications are generally due to either geometric shape or due to over-magnetisation. A constriction in metallic path, through which the magnetic field should pass, causes leakage flux and provides non-relevant indications. Examples of such non-relevant indications are indications at the base of splines, indications at the inside corners of sharp fillets, indications at the root of threads etc.

Leakage flux is also present at sharp edges and at the ends of the components, which have been over-magnetised. Presence of such leakage flux results in deposit pattern which may be called non-relevant indication. Non-relevant indications can be separated from relevant indications keeping the following facts in mind:

(a) If same method of magnetisation is used, indications appear on all the components at same location.

(b) Indications can easily be related to some geometric features such as edges, sharp corners, roots of threads, constriction in the metal path etc.

(c) An experienced operator knows where to expect an indication (relevant one) and thus he can eliminate the irrelevant ones.

4.15 PHYSICAL PROPERTIES DETERMINATION

Physical properties such as hardness, ultimate tensile strength, composition etc. of magnetic materials can be determined using magnetic test principle. Generally, an energising coil is employed to produce an alternating magnetic field in the sample under test. A second coil is placed adjacent to the energising coil and the voltage induced in the second coil provides the requisite information about the desired physical property. In general, the induced voltage from a standard specimen is compared with the induced voltage from the sample under test. Difference in the two measured values of voltage, provides a parameter for evaluating the difference in the magnetic properties of the standard specimen and the sample under test. Either a simple meter may be used to quantify the difference in voltage or a Cathode Ray Oscilloscope (CRO) may be used for evaluating the difference in the two voltage signals. Oscilloscope presentation has an edge over the meter type system because the oscilloscope presentation not only provides the amplitude of the harmonics but also the harmonic content.

Sometimes, two materials with different magnetic properties do not produce any difference in the induced voltage. This happens when the difference in the values of magnetic permeability is counter balanced by the difference in the values of magnetic hysteresis of the two samples. Hence, one should be careful while interpreting the obtained signals.

4.16 RESEARCH TECHNIQUES USING MAGNETIC PARTICLE METHOD

A special type test coil system is sometimes used for the measurement of leakage-flux. Before using the probe; the specimen is magnetised and brought into close contact with a rubber like band impregnated with iron-oxide powder. This band thus records the magnetisation pattern. The recorded band is then scanned with a probe called Forster Probe and the induced leakage flux is reproduced quantitatively on the screen of an oscilloscope. This recorded band can be reused by passing it through an erasing system, just as is done in case of tape-recorders used for listening to music of choice.

Forster probe is also used for sorting different types of steel, *i.e.,* according to their composition. For this purpose, a permanent bar magnet is pressed against the surface of test object and then removed to a safe distance to avoid any subsequent influence of this permanent magnet on test results. Using Forster probe and a suitable meter, the residual strength of the magnetic field is measured and calibrated against the composition of different steels.

There is another research technique known as magnetographic technique. In this technique, so called "magnetic rubber" is used. The technique is especially useful in applications where cracks are present on the inner walls of pipes, tubes etc. which are not accessible easily. These test objects are magnetised by adopting usual magnetisation techniques and then filled with magnetic rubber. Due to leakage flux present at places where the cracks are present on the internal wall; crack pattern is formed on the magnetic rubber and can easily be analysed. This pattern on the magnetic rubber is like a thumb impression and acts as a permanent record too.

CHAPTER 5

Liquid Penetrant Inspection

Liquid penetrant inspection technique is one of the simplest and widely used non-destructive testing technique. Using this technique, one can detect only surface flaws in every materials. The shape and size of component under inspection does not matter as long as the criterion of flaw being open to the surface is met. This condition is a must to enable the penetrant to penetrate into the flaw by capillary action. Depending upon whether a dye or a fluorescent liquid is used as penetrating medium, the technique is called either dye-penetrant technique or fluorescent-penetrant technique.

5.1 HISTORICAL BACKGROUND

As far back as 1920, liquid penetrant inspection technique was used for detecting surface cracks. However, it was in the form of oil and chalk dust method (see Section 5.2), which is crude form of present day's liquid penetrant inspection technique. The oil and chalk-dust method was widely used to inspect railroad axles in railroad shops. Sometimes, oil is used to be heated to lower its viscosity and surface tension to suit a particular requirement. Alternatively, the product under inspection used to be slightly heated to expand the crack and to make its detection easier.

To improve upon the chalk and dust method, oil-vapour blast technique was employed in the past. In this technique, after cleaning the surface of product, the product used to be soaked in kerosene for a few minutes. After this soaking operation, product's surface used to be blasted with clean grit having size around 100 mesh. Thereafter, the component used to be washed with water for the removal of any extra grit and the product was thereafter allowed to air-dry. Indications used to appear on the vapour blasted surface due to the seepage of kerosene from the cracks. These seepage indications were, however, generally very faint and also they used to disappear in a relatively short period of time because of vapourisation of kerosene.

Just like slightly heating the component's surface to expand the surface cracks, one may take help of etching action also to enlarge any tight surface crack. For etching purposes, one may use either dilute hydrochloric acid or sulphuric acid.

5.2 OIL AND CHALK-DUST METHOD

The oil and chalk-dust method may be thought of as liquid penetrant inspection technique in its infancy. It was an age-old practice to dip the component under inspection in a suitable oil bath or in kerosene for some time and then gently wipe off the oil or kerosene from the surface. After the surface used to dry off, chalk-dust (calcium-carbonate or alike) used to be sprinkled all over the surface. Dry chalk-dust layer over the surface of the component, used to suck the remaining oil or kerosene which had earlier penetrated into the cracks. Hence, an observer could detect cracks by looking for wet areas on the chalk-dust layer. Wet areas used to have appreciable reduction in the whiteness.

This oil and chalk-dust method, though very simple in application, had its own limitations. In the present liquid penetrant inspection technique, the shortcomings of oil and chalk-dust method have been removed.

The first and foremost shortcoming of oil and chalk-dust method was that there was not enough colour contrast between the wet-chalk-dust and dry-chalk-dust areas. Hence, an observer searching for surface cracks was liable to miss useful indications. This poor colour contrast problem has now been overcome by making use of red colour dye instead of oil or kerosene. The red colour dye penetrant provides an excellent contrast on white background. The second shortcoming of the oil and chalk-dust method was that the oil or kerosene used did not possess proper viscosity. Sometimes, due to poor viscosity, the penetrant used to flow out by itself and not by sucking action of chalk-dust. This problem of improper viscosity too has now been overcome and the dye used presently are neither too viscous to cause penetration problem into deep sharp cracks, nor the dyes are too thin to be removed during "gentle wiping" operation. Other problems such as uneven spread of chalk-dust resulting in smudging, lost indications etc. have also been overcome by the development of proper quality developing powder and by the development of proper spraying techniques such as spray-guns, aerosol cans etc.

Commercially available dye-penetrant and developers have different mixtures depending upon the make. However, a satisfactory dye-penetrant may be made by mixing red dye and diesel oil. For preparing developer, one may dissolve talc in alcohol. Commercially available cleaners for cleaning the surface under inspection contain suitable liquid solvent such as acetone, carbon tetrachloride, poly-vinyl alcohol etc. Besides liquid solvents, one may adopt vapour blasting, vapour degreasing or acid etching for preparing the surface for subsequent liquid penetrant inspection. Sand blasting should not be used for cleaning the surface because it results in closing up of small surface openings.

5.3 PRINCIPLE OF LIQUID PENETRANT METHOD

The principle of liquid penetrant inspection technique is based on the phenomenon of capillary action. If a component is dipped into a liquid or if a liquid is sprayed over the component's surface, a part of the liquid penetrates into the crack by capillary action. The part of liquid which has penetrated into the crack remains there till it is brought back to the surface by capillary action again (i.e., if one wipes-off the liquid gently from the component's surface; only the liquid on the component's surface shall be wiped-off and the liquid which has penetrated into the crack, shall remain at its place). On application of developer powder (which is nothing but an improved version of chalk-dust or say a sort of talcum powder), the action of which is similar to that of a blotting paper, the penetrant from the crack is sucked back to the surface, again by capillary action. As in the case of blotting paper, the size of indication obtained is larger than the actual crack size. Figure 5.1 diagrammatically represents the principle involved in liquid penetrant flaw detection technique. The figure provides a schematic representation

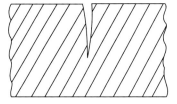

(*a*) Sketch showing a deep sharp crack open to surface

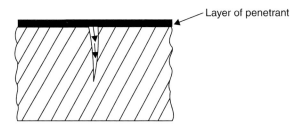

(*b*) Liquid penetrant penetrating into crack by capillary action

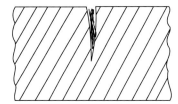

(*c*) On wiping off the penetrant from the surface, penetrant which has
penetrated into the crack remains where it is

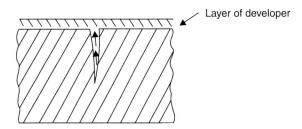

(*d*) On application of developer, penetrant being sucked back, again by capillary action and
due to blotting paper property of the developer, providing the required indication on the surface

Fig. 5.1 Schematic representation of the principle of liquid penetrant inspection technique

of as to how does the penetrant penetrates into deep sharp cracks and as to how it is brought back to the surface on the application of developer. Also, as stated earlier, due to blotting-paper like property of the developer, the size of indication obtained on the surface is larger than the actual crack size and can thus easily be observed and located. The enhanced size of indication helps in observation, in addition to the fact that red coloured dye provides red coloured indication, which has an excellent contrast with white background of developer powder.

As mentioned for oil and chalk-dust method, if the surface of the specimen is warm during the application of penetrant, better indications are obtained. This is due to the fact that heating results in slight expansion of the crack openings and the penetrant penetrates more readily. One may increase the rate of penetration also by slightly raising the temperature of the penetrant. However, before heating the penetrant, one should make sure that it is not going to affect the viscosity of the penetrant adversely and also one should remember that these penetrants have low flash point.

Sometimes, specimens under inspection are heated and then immersed in cooler penetrant. This drop in the temperature causes a low pressure area inside the defect and the pressure differential aids in sucking the penetrant into the defects. One may sometimes achieve better penetration by striking or vibrating the test specimen because it also results in opening-up of the defects for better penetration.

5.4 INSPECTION TECHNIQUE

Liquid penetrant inspection technique is described in the following step-wise fashion:

(*i*) **Cleaning of the surface:** The first step involved in liquid penetrant inspection technique is to thoroughly clean the surface under inspection. This cleaning operation is very essential and improper cleaning may result in spurious indications as well as in missing a few genuine defects. There should not be any dust, dirt, rust, grease, lubricating oil, paint, coating etc. on the surface of component under inspection. Rust, dust, grease etc. generally block the surface-openings of existing cracks and for want of the openings, the liquid penetrant does not penetrate into the cracks. Thus the detection of these cracks is missed. Sometimes, these unwanted items (rust, grease, lubricating oil etc.) not only block the entrance to the surface cracks but they provide spurious indications too. For example, when penetrant is sprayed on a surface having traces of grease on it, the penetrant adheres to grease and it keeps doing so during "gentle wiping-off" operation too. This results in spurious indications of hair line surface cracks, when developer is applied to the surface. Therefore, in order to eliminate spurious indications and to avoid missing the genuine defects, one must thoroughly clean the surface of the component under inspection.

For cleaning small components, first of all, the component is cleaned with a rag (*i.e.,* a piece of soft-clean cloth or alike) to remove dirt/dust, lubricating oil/grease etc. from the surface. Thereafter, wire-brushes are used for rough cleaning. These wire brushes loosen and remove coarse particles of rust/dust etc. from the surface. Generally, these wire-brushes are supplied alongwith the liquid penetrant inspection field kit by the manufacturers of these kits. Cleaning with wire-brushes is followed by further cleaning using fine-brushes (like the one used by painters). These fine brushes, remove the fine particles lying loose on the surface as a result of previous rough cleaning operation. Fine cleaning is further supplemented with final cleaning using "cleaner" fluid. "Cleaner" cans come in the form of aerosol cans with the liquid penetrant inspection field kit and are marked "cleaner" as such. This cleaning fluid is generally acetone or carbon tetrachloride based fluid and is highly volatile too. On pressing the top, a jet of cleaning fluid gushes out of the nozzle of the can and strikes the surface to be cleaned. As this cleaning fluid is highly volatile fluid, the surface to be cleaned dries-off very rapidly allowing the inspector to carry-on to the next stage of inspection more or less immediately after this final cleaning operation.

For large components however, cleaning with rag, wire-brush, fine brush and 'cleaner' fluid in aerosol cans, becomes impractical and they are generally cleaned by dipping into a series of cleaning tanks containing hot water, soap solution, detergents, cold water and chemical solutions (acetone, carbon tetrachloride etc.). Sometimes, if the situation so demands, the components undergo shot blasting or shotpeening operation to remove scales/rust from their surfaces, before they are sent to the cleaning tanks. Suitable arrangements are provided for either agitating the cleaning fluids in different tanks or for dipping and lifting the part under inspection at frequent intervals. This helps in proper cleaning of surface under inspection.

As mentioned earlier in section 5.2; sand blasting should not be used for cleaning the surface because it has a tendency to close the small surface openings.

Surface treatment involving use of strong acid or caustic soda tends to reduce the fluorescing characteristics of fluorescent penetrants. Therefore, such surface treatments should be carried only after fluorescent penetrant inspection (see section 5.9) process in over.

Finally, buffing and certain other surface finishing operations tend to bridge over the flaws. Such operations should therefore be performed only after liquid penetrant inspection.

(*ii*) **Applying penetrant to the surface:** The second step in liquid penetrant inspection consists of applying penetrant to the surface under inspection. Two types of penetrant, *viz.*, dye-penetrant and fluorescent-penetrant may be used. The dye penetrants fall into three main groups, *viz.*, (*i*) spirit soluble, which is removed by organic (chemical) solvents, (*ii*) Pre-emulsified, which contains emulsifying agents making it water-washable and (*iii*) Post-emulsified, in which an emulsifying agent is applied to the penetrant (after application of penetrant to the surface) making it water-washable. If a dye is used as penetrating fluid, the technique is called dye-penetrant inspection technique and if a fluorescent liquid is used as penetrating fluid, the technique is called fluorescent-penetrant inspection technique. Generally red colour dye is used for dye-penetrant inspection technique. Fluorescent penetrant is greenish-yellow in colour and a fluorescent liquid fluoresces under ultraviolet light (or so-called black-light) source. The viscosity of liquid penetrant is carefully selected because too viscous a liquid shall not penetrate the deep-sharp cracks properly and too thin a liquid shall not stay in the crack resulting in flowing out of the liquid by itself from the crack as the time passes, or during handling or movement of the component, or during 'gentle-wiping' operation of the surface ('gentle-wiping' operation is third step in liquid penetrant inspection and has been described later).

For small components, best results are obtained by "spraying" the penetrant all over the surface rather than applying the penetrant to the surface by means of a paint brush or by dipping the component in 'penetrant' liquid. Spraying action provides uniform spread of penetrant all over the surface under inspection. Spraying of penetrant is done either by using a spray-gun or by using aerosol cans containing 'penetrant'. Liquid penetrant inspection field kit contains aerosol can marked 'penetrant' has liquid penetrant, and one can spray the penetrant by just pressing the top nozzle to get a uniform coat of penetrant over the surface under inspection. The nozzle should be kept at a distance of about 200 mm to 300 mm from the surface of component under inspection. This is required in order to avoid smudging of penetrant on the surface and to obtain a uniform coating all over the surface.

After the penetrant has been sprayed uniformly all over the surface under inspection, sufficient time (of the order of 5 minutes) should be allowed for the penetrant to penetrate into deep sharp cracks.

For large sized components, for which spraying of the penetrant becomes impractical or uneconomical, dipping of component into penetrant tank is preferred. The component is left dipped into 'penetrant' liquid tank for sufficient time (15 minutes to 30 minutes) to enable the penetrant to properly penetrate into the cracks and to penetrate in sufficient quantity to obtain subsequent development of flaw indication. The exact penetration time is determined only by experimentation for a particular type of product, which is to be inspected. By varying penetration time, the optimum time for a particular size and kind of defect can be determined.

For detecting very fine surface flaws, one may be required to apply the liquid penetrant more than once onto the same surface. Heating procedure to aid better penetration into tight cracks has already been described earlier in section 5.3.

(*iii*) **Removal of excess penetrant from the surface:** After allowing sufficient time for penetrant to penetrate into surface cracks, excess penetrant is removed from component's surface. This is the third step of liquid penetrant inspection procedure.

While removing excess penetrant from the surface, care must be taken to ensure that the penetrant from inside the cracks is not drawn out in the process. This operation of removal of excess penetrant is usually accomplished in two stages. The first stage consists of "gentle-wiping" of the surface with a rag (or with a piece of cloth). The rag used should not have absorbing quality, otherwise it shall draw out the penetrant from within the cracks too while removing excess penetrant from the surface. The wiping operation should not be only gentle but it should be quick too. After wiping-off, the next step (*i.e.,* the second stage) for removal of excess penetrant is carried out, which consists of spraying 'cleaner' fluid on the surface in such a way that it removes the left-over traces of penetrant from the surface only (after gentle-wiping operation, a few traces of penetrant are still left over the surface). The spraying action is achieved by using aerosol 'cleaner' cans and the spraying action should not draw out the penetrant from within the cracks. This condition is easily met when one directs the 'cleaner' fluid jet along the surface of component under inspection. In no case, the cleaner-jet should impinge directly onto the surface. When the cleaner-jet is directed along the surface, only the left-over traces of penetrant from the surface are removed. However, if cleaner-jet impinges directly on to the surface, penetrant from within the cracks too are removed.

If the component under inspection is small and can be handled easily, it should be suitably oriented to achieve the requirement of directing the jet along the surface. For large components too, best results are obtained if the excess penetrant from the surface is removed following the procedure described above and instead of having aerosol 'cleaner' cans, one may spray the cleaning fluid (again along the surface) using spray-guns or alike. Sometimes, however, spraying of cleaning fluid all over the surface of a large component becomes impractical for one reason or the other, and in such cases removal of excess penetrant from the surface is achieved by dipping the component in a tank containing cleaning fluid. The component is taken out of this cleaner-tank quickly to avoid removal of penetrant from within the cracks too. After taking out the component from cleaner tank, the left-over traces of penetrant over the surface are further removed by dipping the component, a few times in rinsing tanks. Again, care is taken to make sure that the penetrant from within the cracks is not drawn out in the rinsing process.

After removal of excess penetrant from the surface of component either by wiping and spraying process or by dipping process, the components surface is allowed to dry-off completely before proceeding to the next step, *i.e.,* application of developer to the surface.

(*iv*) **Applying developer to the surface:** While describing oil and chalk-dust method earlier in the chapter, it was pointed out that the chalk-dust has now been replaced by developer. For inspection of small components, developer comes in the form of aerosol cans and just by pressing the top nozzle of developer can, one can obtain a uniform coating of developer over the surface to be inspected. One should hold the nozzle at a distance of about 200 mm to 300 mm from the surface of component under inspection and move the developer can to and fro to avoid smudging of developer on the surface and to get a uniform coat.

Developer consists of fine chalk powder suspended in a liquid and this liquid evaporates very quickly after being sprayed. The liquid acts as a carrier of chalk powder or talc from inside the cans to the surface under inspection and within a minute or so after spraying, the developer dries off. Developer powder is off-white colour. The action of dry developer powder is very much like that of a blotting paper. A blotting paper is used for drying up ink on a paper, one invariably notices that the blotting paper quickly absorbs the ink and that the ink-mark on the blotting paper is of larger size than the original size of ink spot. Similarly, when developer starts drying-up, it starts absorbing the penetrant and sucks back the penetrant from within the cracks. Furthermore, as is in the case of blotting paper, the indication thus obtained on the surface are of larger size than the actual crack size. The enhanced size indications are easily observed. In case of dye-penetrant testing, the penetrant used is a red colour dye and this red colour dye indication provides an excellent contrast on the white background of developing powder. This colour contrast too, like enhanced size, aids in easy observation of the required indication.

For big size components, spraying of developer using aerosol cans becomes impractical and uneconomical. Therefore, one is required to go for spray-guns or some other suitable means for getting a uniform coat of developing powder on the surface to be inspected. One is required to allow some time for the developer to dry-up to enable it to suck the penetrants from within the cracks, back to the surface.

It is essential to have a uniform coating of developer on the surface because at places where there is too thin a coat of developer, it may not be able to suck-back the penetrant from within the cracks and at places where there is too thick a coat of developer (*i.e.*, smudging on surface), it may result in hidden crack indications. Hence, it is absolutely essential to have a uniform coating of suitable thickness of developer on the surface of the component to obtain proper indications.

Hot air is recommended for quick drying of spray type developers. This may be done using specimen dryers. This hot drying serves a two-fold purpose of quickly drying the developer by evaporating the solvent base and heating of the specimens aids in bringing out the penetrant from the cracks to the surface relatively easily.

(*v*) **Interpretation of indications obtained:** The indications obtained as a result of liquid penetrant inspection do not provide details regarding the exact size of crack (*i.e.*, how deep the crack is or dimensions of the crack etc.). The technique simply provides the location of surface cracks. By observing the indication, an experienced inspector may deduce a rough idea regarding the crack size (*i.e.*, the length and the width of the crack). Deducing information regarding the depth of crack (inside the component), however, poses problem. Figure 5.1 schematically represents various steps involved in liquid penetrant inspection technique.

The coating thickness of developer has a significance of its own and it determines the minimum defect size that is detectable. If the coating thickness is more, small defects are not revealed because of insufficient penetrant to diffuse completely through the developer coat. Dry developer reveals even the finest and smallest surface defects.

5.5 COMMERCIALLY AVAILABLE DYE-PENETRANT INSPECTION KITS

There are several brands of dye-penetrant inspection kit which are commercially available. All these inspection kits generally contain a wire-brush, a fine brush, 4 aerosol cans of cleaner fluid, 2 aerosol cans of dye-penetrant and 2 cans of developer. All the aforementioned items come in handy vanity type wooden box and can easily be taken to the inspection sites.

Fluorescent-penetrant inspection kits are also just like dye-penetrant inspection kits. The only difference being that instead of dye-penetrant aerosol cans, fluorescent kit contains fluorescent-penetrant aerosol cans.

Cleaner fluids, dye-penetrant, fluorescent-penetrant and developer powder are also available in packings other than aerosol cans. However, as pointed out earlier, for best results, one should go for spray method using aerosol cans, as far as possible. For inspection of large components, bulk packings of cleaner fluid, liquid penetrant and developer powder are commercially available.

5.6 INDUSTRIAL APPLICATIONS

As mentioned earlier, using liquid penetrant inspection technique, one can easily detect surface cracks in all types of material. Liquid penetrant inspection technique is in wide use in industry and has several industrial applications. Some of the industrial applications are listed below:

1. Using liquid penetrant inspection technique, incoming materials are inspected for probable surface defects such as forging cracks, rolling defects, casting cracks, surface pits and undesirable markings etc.

2. During machining and finishing operations, certain undesirable and objectionable surface defects do crop-up. These surface defects are quite deleterious especially so if the component is subjected to fatigue loading. These defects are tool marks, grinding cracks, machining cracks, die marks etc. These defects can easily be detected using liquid penetrant flaw detection method.

3. During fabrication processes also, sometimes, surface cracks are unavoidably created. For example, during welding and after welding, sometimes surface cracks result due to uneven thermal expansion and shrinkage. These surface cracks can also easily be detected by liquid penetrant inspection technique.

4. If proper precautions are not taken during heat treatment process, it results in development of surface cracks. This is particularly so during quenching operation. These surface cracks (heat-treatment or quenching cracks) can be located using liquid penetrant inspection technique.

5. In-process and service cracks can also be detected, provided these cracks are surface cracks. Sub-surface and internal service cracks cannot be detected using liquid penetrant inspection technique. As is well known, fatigue cracks mostly initiate from

the surface and propagate inward. For metallic materials, fatigue crack propagation time is usually large. Hence, detection of these surface cracks (fatigue cracks) at initial stages of fatigue crack propagation, during routine or periodic inspection, using liquid-penetrant inspection technique, could avoid the subsequent catastrophic fatigue failure.

6. Generally, if the component under inspection is ferromagnetic, magnetic particle inspection technique is preferred over liquid penetrant inspection technique because magnetic method provides rapid and unambiguous results. However, for non-magnetic materials, liquid penetrant inspection technique is widely used in industry. Sometimes, even for ferromagnetic component, liquid penetrant inspection technique is used for varied reasons.

7. In certain applications, surface finish of a ferromagnetic component is of utmost importance. If the surface of such components is checked for surface defects using magnetic particle inspection technique and if electrical prod method is used to magnetise the component, the prods may spoil the surface finish by leaving spots at the points of prod contact due to arcing. In such cases, liquid penetrant inspection technique is preferred.

8. For fibre composites (*i.e.*, GRP, CFRP etc.), surface cracks are generally absent in newly made composite. However, these cracks may develop during the loading history or during the service life of these fibre composites. Surface cracks in fibre composites, polymeric materials etc. are also revealed by liquid penetrant inspection technique. In case of newly made fibre composites, liquid penetrant inspection also reveals the surface texture (specially so if bleeder cloth is used while making the fibre composite), surface pits etc.

5.7 PRECAUTIONS AND LIMITATIONS

The first and foremost precaution which should be observed is that before the start of liquid penetrant inspection, the surface should be thoroughly cleaned. If the surface is not cleaned thoroughly, it may results in missing a few genuine defects and may also result in a few spurious indications, as has already been described in section 5.4. If the surface is painted, rusted or coated with something which may block the openings of surface cracks; the paint etc. should be removed before the start of inspection procedure.

The inspection procedure described in section 5.4 should be carefully followed. Wrong usage may result in incorrect indications leading to rejection of good components or acceptance of faulty components. The step which needs maximum attention is that while removing the excess penetrant from the surface, care must be taken to make sure that the penetrant from within the cracks is not drawn out in the process. Hence, the wiping operation should be gentle and quick. Also, the cleaner jet should be directed along the surface and should not impinge directly onto the surface.

As far as possible, from time to time, standard test blocks with known surface defects should be inspected using the current batch of liquid penetrant inspection kit to establish, whether or not, the dye-penetrant, developer etc. in the inspection kit are working satisfactorily.

Another major precaution which should be observed is that the test should be conducted in an open area or in a well-ventilated room and not in a congested area or within a close room because vapours arising from the penetrant may be toxic. If there is a restriction on the availability of open space, proper exhaust facility, such as exhaust fans, should be incorporated to remove the unhealthy vapours (due to spraying of cleaner, penetrant and developer) from the test area.

It has been observed that in certain cases, dipping of the component in hot water helps in opening-up a little bit of the too-tight (or so-called closed cracks) surface cracks in which the penetrant could not have penetrated otherwise. So in such cases, dipping of the component in hot water and then drying it with warm air before the application of dye-penetrant should be preferred, provided the hot water dipping does not affect the component otherwise.

The inspection area should be properly illuminated to avoid missing probable feeble indications. The illumination requirement is especially important when the inspector has to inspect many components within a short span of time.

Finally, one must follow the instructions provided by the manufacturers of the penetrants because these penetrants may be highly toxic or may have a low flash point or may be highly volatile. Sometimes, the oil base of the penetrant may cause irritations to the skin. One may protect his skin by wearing proper aprons and gloves. Washing with soap and water generally helps in the removal of penetrant from the skin and clothing.

Regarding the limitations of liquid penetrant inspection technique, the first limitation is that one can detect surface cracks only. The technique does not provide a solution for detection of internal cracks or sub-surface cracks. The technique also fails to detect wide shallow cracks (such as indentation marks of previous hardness test or depressions on surface or alike) because these wide shallow cracks do not possess capillary action and the penetrant from these wide shallow cracks is removed while removing the excess penetrant from the surface. As these wide-shallow cracks are incapable of retaining the penetrant in them, they do not show-up on the application of developer.

Other limitations of liquid penetrant inspection are that a well painted surface cannot be inspected as such and the paint has got to be removed before carrying-on with the inspection. Furthermore, this inspection technique is quite slow in providing results and for every inspection, an appreciable quantity of cleaner/dye/developer etc. is required.

In case of small diameter tubings, liquid penetrant inspection technique fails to detect cracks at the inner surface of the tubings because of the inaccessibility of the interior portion of the tubing. Outer surface of these small diameter tubings may, however, be inspected using penetrant inspection technique.

5.8 TEST BLOCKS

By heating a flat piece of oil quenched tool steel to about 800°C and then by quenching it in water, one may obtain a test block having a large number of surface cracks. Even discarded lathe tools may be heated to about 800°C and quenched subsequently to obtain surface cracks.

One may also make another test block by taking a piece of aluminium (say $100 \times 50 \times 5$ mm) and heating it to a temperature of about 500°C followed by quenching in cold water. This produces a typical crack pattern at the surface of aluminium test block.

Yet another test block containing cracks of different sizes may be made by taking a tapered steel rod having a number of drilled holes along its length. This steel rod can then be loaded on a torsion bench and necessary torque may then be applied to produce cracks. Due to taper of the rod, the shearing stresses tend to be different at different drilled holes. The value of shearing stresses depend upon the position of the hole and on the ratio of the hole diameter to the diameter of the rod at that particular position. Due to different shearing stresses, cracks of different sizes are produced at different holes. This type of test block is very useful for checking the inspection capability of penetrant system from time to time.

5.9 FLUORESCENT PENETRANT TESTING

As already mentioned earlier, liquid penetrant inspection is of two types: (*i*) dye-penetrant testing if a dye is used as a penetrant and (*ii*) fluorescent-penetrant testing if a fluorescent penetrant is used instead of dye.

A fluorescent liquid is greenish yellow in colour and unlike red dye, it does not provide a good contrast with the white background of developer powder, when observed in a well-illuminated room. However, if one observes the same in a dark room and under a black-light source (*i.e.*, safe ultraviolet light source), the liquid and therefore it indication shines brilliantly and can easily be located.

The principle and inspections technique for fluorescent-penetrant is same as that for dye-penetrant method. The only difference being that instead of using dye, one uses a fluorescent liquid as the penetrant and instead of observing the indication in an ordinary room in well-illuminated area, one is required to have a dark room and a black light source for observing the fluorescent penetrant indications.

Hence, if one opts for fluorescent-penetrant inspection instead of dye penetrant inspection, an added investment in form of dark room and black light source is required. However, it is far more easier and quicker to locate a brilliantly shining crack in dark background as compared to dye-penetrant test indications. Hence, when one is required to inspect components on mass scale, fluorescent-penetrant testing has an edge over dye-penetrant testing. Use of fluorescent penetrant permits inspection speed of test indications, as much as six to ten times those possible with test indications of dye-penetrant test.

As already described in Chapter 4; commercial black light sources produce black light (near ultraviolet light) of a specific wavelength which is non-injurious to eyes or skin. Hence, this is called safe black light too. Fluorescent indications fluoresce under black light and appear as brilliantly shining indications.

5.10 DETECTION OF THROUGH LEAKS

Liquid penetrant inspection technique may be used for the detection of through leaks as well. In such an application, the penetrant is applied on one side of the object under inspection and the other side is thoroughly examined for visible indication of red dye or indication of fluorescent liquid, taking help of a black light source. Sometimes, even developer is used on the other side for rapid detection of dye or fluorescent liquid which has leaked across the thickness of the object under inspection.

The penetration time depends on the type of object under test and also upon its thickness. If the through crack is such that the capillary action shall take place rapidly, penetration time shall be low. However, if a coarse porosity is present in the leakage path, it shall reduce the capillary action and penetration time shall increase considerably.

Liquid penetrant flaw detection technique is useful for the detection of through leaks particularly for thin-walled structures, say thickness up to 5 mm. If the through leak has trapped moisture or dirt in the leakage path, they shall restrict the penetrant from penetrating across the thickness and incorrect results may be obtained. Hence, while using liquid penetrant inspection technique for the detection of through leaks, one should carefully interpret the presence or absence of penetrant indications.

5.11 TYPICAL INDICATIONS ASSOCIATED WITH LIQUID PENETRANT TESTING AND THEIR INTERPRETATIONS

Liquid penetrant inspection indications depend on the type of penetrant used. If fluorescent penetrant is used, defect show-up as glowing yellow-green dots or lines against a dark background. In case of dye-penetrant testing, defects are indicated as red dots or lines against a white background.

The interpretation of the characteristic patterns indicating the types of flaws is of great significance. For example, a crack of a small opening is indicated by a line of penetrant. Dots of penetrant indicate pits or porosities on the surface of object under inspection. Such dots appear over an area or as isolated spots and generally do not form a line pattern. A series of dots forming a line pattern indicates a tight crack, cold shut or partially welded lap. Fatigue cracks also generally appear in the form of a series of very fine dots.

One may obtain a rough estimate of the surface opening by measuring the width of the indication (*i.e.,* the amount of spreading of the penetrant on the developer). However, there does not exist a definite relationship between the surface opening and spreading of the penetrant. It is only with experience that an inspector may relate between these two parameters for a particular type of object under inspection.

CHAPTER 6

X-Radiography

6.1 INTRODUCTION

One of the most widely used non-destructive testing technique is X-radiography, in which X-rays are used as detecting medium. Instead of using X-rays, if one uses gamma rays, it becomes gamma radiography. The radiographic techniques work on the principle of differential absorption, *i.e.*, different materials absorb penetrating radiations differently. A denser material like lead would absorb practically all the radiations (say X-rays), whereas a material like air will let all the X-rays pass through without any absorption. Hence, if there happens to be an air pocket entrapped in a casting (*e.g.*, blow holes); X-rays would be absorbed by the casting differentially. This property makes it possible to detect the blow hole by checking the intensity of received X-rays at different locations. X-ray films are commonest detecting medium for this purpose.

X-rays are extensively used for medical diagnostics. Working principle of medical radiography (or Roentgenography) and industrial radiography is same. Medical radiography is usually called Roentgenography to honour Nobel Laureate William Conard Roentgen, who discovered X-rays in 1885 AD. Industrial radiography followed later around the year 1920 AD. In medical radiography, the doctor (or radiologist) can ask the patient "where does it hurt?" and take X-ray picture of only that portion of body. An engineer cannot ask the metal or any other material under inspection as to where the defect is and hence he has to take X-ray picture of whole material which is being inspected to look for defects. However, an engineer is in an advantageous position because he is free to expose the material to X-rays as long as he requires but a doctor cannot expose the patient to X-rays even for a second. Mostly exposure timings in medical radiography are only a fraction of a second. This limitation is due to the fact that X-rays are injurious to body tissues and cells. Longer exposure timings kill these cells and may cause burns and permanent damage.

6.2 PROPERTIES AND PRODUCTION OF X-RAYS

X-rays possess the following properties:

     1. Are invisible high frequency electromagnetic radiation (Frequency 1018 Hz or higher).

     2. Can penetrate matter, which are opaque to light.

     3. Are differentially absorbed.

4. Travel in straight lines.

5. Produce photochemical effects in photographic emulsions.

6. Ionize gases through which they pass.

7. Are not affected by electric or magnetic fields.

In a conventional X-ray tube used in radiographic equipments, X-rays are produced when high speed electrons are made to collide with a tungsten target. The X-rays produced are so called white X-rays and contain all possible wavelengths from very long down to a short wave minimum or cut off wavelength λc in Å is given by

$$\lambda c = 12.4/V$$

where   V = accelerating voltage in kilovolts applied to X-ray tube. Thus if an X-ray tube runs at 100 kV, it cannot produce radiation of wavelength smaller than 0.12 Å.

Lower wavelength X-rays have more penetration ability and are less scattered (less fogging of X-ray films). However, long wavelength X-rays provide better contrast making defect resolution easier. Hence, one has to select a wavelength to optimise between penetration depth, fogging and contrast.

Since the penetrating ability depends upon the wavelength, one may expect radiographic equipment to have a knob to set the wavelength in nanometers at a particular level. This is, however, not so and radiographers adjust the voltage setting (kilovolts knobs) to obtain proper penetration. It may be mentioned here that the penetrating ability of X-rays does not depend on flux produced by X-ray generators. Higher flux only makes it possible to expose the X-ray film quickly.

Practically all materials absorb X-rays, *i.e.,* they attenuate X-rays passing through them. This attenuation of X-ray beam depends upon the mass absorption coefficient of the medium (*e.g.,* material under inspection) through which the X-rays are passing and it also depends on the thickness of the material under test. Denser the material, more is attenuation of X-ray beam. Similarly, thicker the material more is the attenuation of X-ray beam. Attenuated X-ray beam is less intense than the original X-ray beam coming out through the window of X-ray equipment. The exact relationship between the intensities of the transmitted (attenuated) beam and the incident beam is as given below:

$$I_t = I_0 e^{-\mu t} \qquad \qquad \qquad ...(1)$$

where I_t = intensity of transmitted (attenuated) beam

I_0 = intensity of incident (original) beam

μ = mass absorption coefficient of the material

t = thickness of the material under test

The mass absorption coefficient (μ) depends upon the density of the material as well as on the wavelength of X-rays being produced for inspection.

X-rays are generated using a vacuum tube device containing a tungsten target (which is embedded in copper), a source of electron and a means of accelerating these electrons. When high velocity electrons impinge on the metallic target, X-rays are produced. X-ray units generally have Beryllium window, which can be positioned at any angle to suit the inspection process. X-ray tubes require 10 to 400 kV operating voltage and this high voltage is generated by a small transformer in the tube head.

6.3 WORKING PRINCIPLE OF X-RADIOGRAPHY

As already described earlier, the X-radiography NDT technique works on the principle of differential absorption. It is explained in Figs. 6.1 and 6.2 diagramatically. In Fig. 6.1, a material which contains a defect and which is having a mass absorption coefficient (μ) is placed under an X-ray source. Since air does not attenuate the X-ray beam, one may assume that the defect does not absorb X-rays or (for defect may be taken as zero, as shown in Fig. 6.1). Hence, the intensity of the X-ray beam as detected by any form of detector located at A will be lower than that recorded at B, because the path to detector A contains a greater amount of absorbing material. The intensity of the X-ray beam reaching at point A will be governed by the equation 1 given earlier and one may write the following expression for the intensity of X-ray beam reaching at point A.

$$I_A = I_0\,e^{-\mu\,t} \qquad\qquad\qquad ...(2)$$

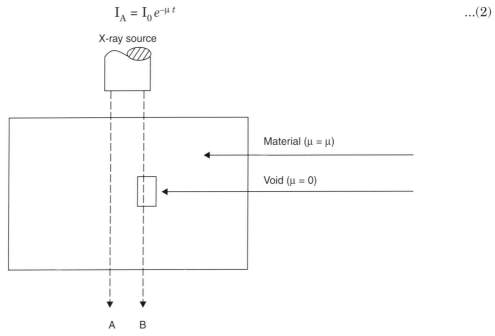

Fig. 6.1 Principle of differential absorption of X-rays

The change in intensities at points A and B can be expressed by ΔI and it can be obtained using the following expression:

$$\Delta I = \frac{d\,I_A}{dt}\,\Delta t \qquad\qquad\qquad ...(3)$$

where Δt is the thickness of defect measured parallel to the thickness of material or so to say measured parallel to the direction of X-ray beam travel. To solve equation (3), one can differentiate equation (2) first and the value may then be substituted in equation (3) as shown below:

$$\Delta I = -\,\mu I_0 e^{-\mu\,t}\,\Delta t$$

or $$\qquad\qquad \Delta I = -\,\mu I_A \Delta t \qquad\qquad\qquad ...(4)$$

The detection sensitivity is denoted by letter S and is given by the change in the beam intensity produced by the defect compared to the intensity without defect. Mathematically,

$$S = \frac{\Delta I}{I_A} = -\mu\, \Delta t \qquad\qquad ...(5)$$

It is evident from equation (5) that for maximum senstivity (S_{max}); μ, should be maximum. As μ is material dependent as well as dependent on the wavelength of the X-rays; sensitivity can be increased by selecting a proper voltage. Higher the wavelength, larger is the value of μ. One cannot change the material to obtain higher μ value because material under inspection is already fixed and that material will have only a particular mass absorption coefficient for X-rays of a specific wavelength.

Figure 6.2 shows a specimen with varying thickness and containing a high density inclusion as well as a porosity or void ($\mu = 0$). This specimen has been placed under an X-ray source emitting diverging X-rays. A photographic film (X-ray film) has been placed below the specimen to record the modulated X-ray beam intensity. Just as in case of photographic film, X-ray film will become darker in the area where there is more exposure and the film will be

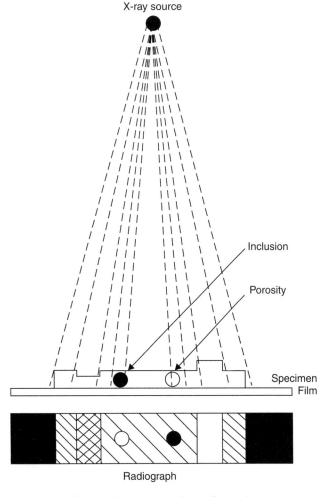

Fig. 6.2 Production of a radiography

lighter (or greyer) in areas where there has been less exposure. Hence, a high density inclusion, which absorbs all the incident X-rays and does not let any X-ray reach the X-ray film, will be shown as a bright spot as shown in Fig. 6.2. A porosity or a void which is nothing but air entrapped and which has $\mu = 0$, will let practically all the X-rays pass through it. Hence, the X-ray film below such a spot will be highly exposed and the porosity will appear as a dark spot on the X-ray film. Besides inclusions and porosities, thickness too alters the exposure level. Hence wherever the thickness is least, shade will be darker compared to the shade for the portion having more thickness, as shown in Fig. 6.2. The portion of the X-ray film, which receives uninterrupted X-rays (*i.e.,* the portion not covered by the object under test), gets fully exposed and becomes completely dark. The developed X-ray film is called radiograph and is generally viewed against bright light source or so-called X-radiograph viewers. Methods adopted for developing and fixing of X-ray film or plate is on same lines as that used for photographic film. However, positive prints which are obtained for photographic films are not required in case of X-radiography and the developed X-ray film (or so-called "negative") itself serves the purpose of inspection and permanent record of inspection.

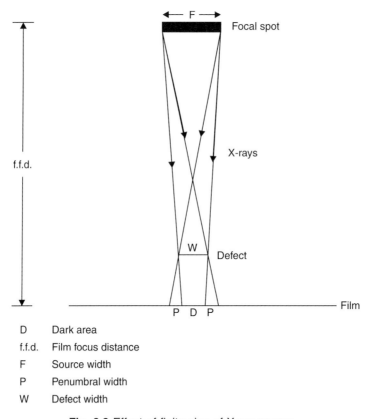

D	Dark area
f.f.d.	Film focus distance
F	Source width
P	Penumbral width
W	Defect width

Fig. 6.3 Effect of finite size of X-ray source

Figure 6.3 explains certain terms which are generally used in connection with X-radiography. For example, the distance between X-ray source (window) and the radiographic film is called f.f.d. or film to focus distance as shown. Similarly, as the X-ray source is not a point source, a focal width (F) which is also called source width is associated with X-radiographic equipments. The source width is reasonably small but it certainly cannot be called a point source. These two parameters, *i.e.*, f.f.d. and F, govern the sharpness of defect image, as explained in Fig. 6.3. As a guide, an acceptable penumbral width for studying a radiograph for defect with unaided eye should be about 0.25 mm. Penumbral width and the dark area showing defect have been illustrated in Fig. 6.3. Experience tells that for sharp image of the defects, f.f.d. should be large (around 500 mm) and the distance of defect from film should be small, *i.e.*, as close as practically possible.

The X-ray film, which is placed below the object under inspection receives both the primary and secondary beams. Primary beam refers to X-ray beam which travels in a straight line and reaches X-ray film after travelling through the object under inspection. The secondary beam refers to those beams which are scattered beams. These secondary beams are due to scattering taking place at the time when the primary beam strikes the object under inspection or due to scattering of reflected X-ray beams from the surroundings and from within the material. The secondary beams produce a uniform darkening of the film, both within and outside the specimen area. This uniform darkening or "fog" reduces contrast of radiograph and defect resolution becomes difficult. In extreme cases, darkening due to fog may be so much more than the darkening due to primary beam that even the specimen is not resolved on the radiograph, what to say about the defect within the object under inspection. To avoid fogging wavelength of the X-rays used should be short or one should go for intensifying screens. These intensifying screens are placed next to the radiographic film to absorb scattered X-rays and thereby enhancing darkening due to primary beams. At voltages higher than 30 kV, long wavelength X-rays are produced necessitating usage of intensifying screens.

6.4 VARIOUS METHODS FOR DETECTING MODULATED INTENSITY OF X-RAYS BEAM

When X-ray beam passes through an object, its intensity gets modified or modulated and one has to check the distribution of modified beam intensity to evaluate flaws or inhomogenities in the material. For this purpose, one may use photographic film, fluorescent screens, ionisation of gas, counters and scintillation crystals or some electronic devices such as a radioscope. In computerised tomography too, electronic devices are used for checking the modulation in the beam intensities. However, as in case of industrial radiography, one requires a large detector surface which can give information about the distribution pattern of modified X-ray beam intensity; ionisation of gas and counters etc. do not provide a satisfactory solution. Hence, generally for industrial radiography either film or fluorescent screens are used. Out of these two, photographic emulsion (film) has some definite plus points as compared to fluorescent screen. These plus points are as listed below:

(*i*) As the film integrates the received radiation over the period of exposure, even very weak X-ray intensities can be used by employing longer exposure time.

(*ii*) X-ray films of varying speed are available and, therefore, by sacrificing speed, one may go for fine grain film. Use of fine grain film is very helpful in applications such as microradiography, where radiograph is taken using fine grain film and then enlarged to get the finer details of the portion of interest.

(*iii*) One can get different contrast conditions by selecting different working film densities.

Fluorescent screens are used to monitor the changes in the intensity of transmitted X-ray beams. In all imaging techniques with X-ray beams, one has to capture the X-ray shadow (*i.e.,* distribution of transmitted X-ray beam intensity) and process them using different sensors to produce real time images. Fluorescent screens make it possible to see the radiograph on the screen without going for development of X-ray film. However, a permanent record of radiographic inspection is not possible when the inspector uses fluorescent screen. In fluorescent method of radiography or in fluoroscopy, the object to be inspected is moved under the X-rays source to search for defects along the length of the object. The object can be rotated too, if required, while looking for a defect at an angle. Best image of defects is obtained when the defect is aligned along the direction of X-ray beam travel. However, fluoroscopy is not as sensitive as film radiography and therefore film radiography is more widely used.

Using X-ray tomography, one can obtain X-ray picture of a particular cross-section of any material. This is achieved by blurring of images from the nearby planes, leaving only the image of the section of interest. This is achieved by moving the X-ray source and the detector in such a way that only the shadow of the plane of interest remains in focus and all other images are out of focus. In computed tomography, image is produced by computer techniques using the data of transmitted beam intensity as the detector, the source and the object relatively moving past each other.

6.5 APPLICATIONS OF X-RADIOGRAPHIC TECHNIQUE

X-radiography is one of the oldest NDT technique for checking the integrity of various engineering materials. Unlike magnetic methods of crack detection where the material under inspection has to be magnetic material (ferromagnetic); radiographic technique is applicable to all types of engineering materials, *i.e.,* metals, polymers, ceramics, composites etc. Also, unlike liquid penetrant inspection which can be used only for inspecting surface defects; X-radiography can be used for detection of surface defects, sub-surface defects or internal defects. However, defects which are along the direction of X-ray beam travel are more readily detected. If the defect is very thin (such as a hair line crack) and is perpendicular to the direction of X-ray beam travel, it may be missed because such defects have hardly any dimension along the direction of X-ray beam travel. In FRP composites, translaminar cracks which are along the direction of X-ray beam travel are easily detected but inplaner crack or inplaner voids are missed by if one opts for X-radiographic NDT technique. Fibre wash and orientation of fibres can also be inspected in FRP composites using X-radiography technique.

Besides being extensively used for checking defects in incoming materials such as plates, pipes etc. X-radiographic technique is also used for inspecting structures, devices, components, etc. Blow-holes in castings and welding defects such as lack of penetration, heat-affected zones, cracks etc. are easily inspected using X-radiography. Inclusions in tubes and corrosions pits in tubes are also detected using radiographic technique. Sometimes, radiography is used in conjunction with some other NDT techniques too. For example, radiography supplemented with liquid penetrant inspection or radiography in conjunction with ultrasonics etc. These combinations help in detecting those flaws also which could not have been detected using X-radiographic technique alone.

Radiography technique can also be used for signature pattern of archaeological statues. For example, a bronze statue hailing from bronze-age may be radiographed and its radiograph would be unique to that particular statue only. Any fake or replica of statue will give a different radiograph because any internal blow-hole, pit or dent in the original statue cannot be built into the replica in exactly same way. This will help in distinguishing between the original and the fake statues.

In the field of biomedical engineering, X-radiography is used for the inspection of implants, prostheses, orthopaedic and orthrotic devices etc. for checking their quality and structural integrity.

6.6 SAFETY ASPECTS RELATED TO X-RADIOGRAPHIC TESTING

While using X-radiographic technique for inspection of materials or structures, one has to be careful and take certain necessary precautions. X-rays production has helped us in locating internal defects in castings etc., but unfortunately X-rays ionise biological tissues and are therefore a potential health hazard if proper precautions are not taken by the X-radiography personnel. The following precautions need to be taken while opting for X-radiographic inspection:

(i) X-radiography test should be conducted in an area which is away from crowded area and out of people's movement path ways.

(ii) If the industry is working in three shifts, X-radiography test should be done in the night shift when number of workers are minimum.

(iii) The test area should be properly identified and marked "X-rays" or "Danger X-ray Testing Area" or "Keep-out-Radiation Area" to discourage free movement of people in the test area. Only authorised personnel should be allowed in the test area.

(iv) The cabin or the room where the X-ray machine is installed should have brick or concrete walls to avoid X-rays leaving the room. X-rays can penetrate plywood or hardboard partitions easily. Lead sheets absorb X-rays and do not allow X-rays to pass through them. Hence, if lead sheets can cover the inside wall of the cabin or the room; best protection would be obtained.

(v) Person operating the X-ray machine should wear lead apron, lead gloves, protective glasses etc. and should be as far away from the X-ray machine, as possible while the machine is "on", i.e., when the test is in progress. Ideally, the machine should be operable using a remote switch from outside the room.

(vi) X-radiographic inspection area should be surrounded by lead-sheets, if possible. This helps in absorption of reflected and scattered X-rays. These reflected and scattered X-rays are present because whenever a test object is placed under X-rays, a major portion gets transmitted through the object to expose the detecting medium such as X-rays film. However, as the X-rays strike the top surface of the object under test; a portion of these incident X-rays get reflected and/or scattered. As mentioned earlier, these reflected and scattered X-rays cause fogging and they are injurious to any one who may be standing near the X-ray machine. The screens used to surround X-ray test area are called intensifying screen.

Finally, one must remember that excessive exposure to X-rays may cause decrease in the White Blood Corpuscles (WBC) count in blood and may cause leukopenia, anemia and low blood pressure. Severe exposure can result even in skin burns which may prove fatal. X-ray personnel who do not take proper safety precautions develop skin diseases and cancer. Hence, to get a warning about the exposure level, one must wear X-ray dosimeter or a radiation monitoring badge.

The badge contains a small X-ray film (dental X-ray film) in a plastic housing. Analysis of the film in the badge is done regularly every month by a central radiation protection agency. The result of the analysis are communicated to the person directly and if there is a case of overexposure, the employer is also informed about the same to enable the employer to shift that particular person (*i.e.,* who has got an overdose of radiation) to an assignment which does not involve exposure to radiation.

6.7 CONCLUDING REMARKS

X-ray films are generally polyester based and the photographic emulsion is coated on both sides of film. Hence, it does not matter as to which side is facing up. For thin laminates of fibre reinforced plastic composites, good radiographs can be obtained by putting the FRP laminate on top of a black polythene bag containing X-ray film or putting both the FRP laminate and the film together in a light proof black polythene bag. Of course, laminate would be on top of the X-ray film. However, for metallic specimens, X-ray film cassettes are used. These cassettes hold the X-ray film in a metallic enclosure (a thin box like device) which is light proof and which contains a potassium-iodide sheet inside the top cover portion of the cassette. When X-rays fall on this potassium-iodide sheet (which in contact with the X-ray film), crystals of potassium-iodide covert X-rays into light rays which in turn expose the film differently depending upon its intensity. In conventional radiography, it has been noticed that X-radiographs obtained by using film cassettes turn out to be of better quality as compared to the radiographs which are taken without the aid of film cassettes.

Finally, to facilitate detection of flaws in materials, high density fillers such as Conray 420 or barium sulphate can be used, provided these defects have an opening at the surface from where these high density fillers or dye can be injected into the defects. Being high density fillers, they absorb X-rays and produce light areas on the radiograph and easy evaluation of defects becomes possible. X-radiography using high density fillers is a very effective way of studying crack path in FRP cylinders or pipes which may have undergone burst-pressure tests because these high density fillers are in liquid form and the liquid spreads uniformly throughout the delaminated portion or crack easily on being injected.

CHAPTER 7

Acoustic Emission Testing and Acousto-Ultrasonic Testing

7.1 INTRODUCTION

Whenever we tear a paper, break a twig or chalk, bend-unbend a tin sheet or foil, crack a timber or alike, we do hear the tearing or breaking processes. In other words, acoustic (sound) energy is released in these processes. Now sound is a form of energy and we all know that energy cannot be produced or destroyed. We can only convert one form of energy into another form. Hence, one should ponder as to which form of energy is getting transformed into acoustic energy in these tearing/breaking processes? It is something like this. Every material has stored elastic energy within the volume of the material and it is this stored elastic energy that gets released in the form of acoustic energy whenever material undergoes plastic deformation, phase transformation or fracture.

7.1.1 Definition

Acoustic Emission may be defined as elastic waves spontaneously generated (within the volume of material) due to the release of stored elastic energy of the material as it undergoes plastic deformation, phase transformation or fracture.

7.1.2 Synonyms

Various authors in the past and some of the authors even now use terms other than acoustic emission to describe the phenomenon stated above. Commonest of all is Stress Wave Emission (SWE). However, terms such as Stress Wave Analysis Technique, elastic radiation, clicks, elastic shocks etc., have also been used in past.

7.2 NEED FOR DETECTING EQUIPMENT

Whereas tearing of paper or breaking of twigs, twinning of tins (also known as 'tin cry'), cracking of a timber, cracking of rocks etc. produce acoustic emissions in the audible range, most other acoustic emissions are either too low in amplitude or too high in frequency (*i.e.*, ultrasonic range) to be detected by the unaided ear and, therefore, we do require the use of proper detecting sensors and equipments.

7.3 HISTORICAL BACKGROUND

Sound and stress waves generated in various materials have been noted in the past for several years. However, scientist Joseph Kaiser was the first to detect and systematically process the slight sounds produced by the deformation of certain metals such as zinc, steel, copper and lead.

One of the observations made by the Kaiser was that acoustic emission phenomenon is irreversible, *i.e.,* acoustic emissions are not generated during the reloading of a material until the new stress level exceeds its previous stress level value. This phenomenon is widely referred to in the literature as "Kaiser effect".

Researchers in the field of acoustic emissions, however, expressed some doubts to the general validity of Kaiser effect especially for less homogeneous and anisotropic materials such as unidirectional fibre reinforced plastic composites. Also in cases of fatigue and fracture, during the later stages of crack propagation/growth, plastic deformation continues without any further increase in the applied load and acoustic emissions continue. Hence, Kaiser effect is not operative in case of such similar situations. The author also tried to verify the validity of Kaiser effect and for this purpose, a number of carbon fibre reinforced plastic composite specimens were tested using acoustic emission technique. It was observed and confirmed that the Kaiser effect was not fully operative. Repeat of the tests invariably gave a lower count but not zero count as would be expected if Kaiser effect was fully operative. On further investigation it was noticed that this Kaiser effect violation might be because insufficient time may have been permitted for all the counts to appear. The same is explained in the following paragraph.

If the load were held until all acoustic emissions emanating from the sample undergoing loading schedule stop and only thereafter we unload and reload the sample, emission would not occur until the previously attained load is exceeded. However, if the load is not held at the same level for sufficiently long duration and without giving sufficient opportunity for all the counts to surface, we decide to go for unloading and reloading, emissions which did not get sufficient opportunity to surface, would occur at a load that is lower than the one which was attained previously. This phenomenon is generally referred to as modified Kaiser Effect or Felicity Effect. The ratio of load at which emissions reoccur compared to the previously attained maximum load is referred to as Felicity Ratio or Felicity Percentage. Felicity ratio is usually indicative of the amount of damage that has crept into the material due to loading schedule being followed.

7.4 BASIC PRINCIPLE OF ACOUSTIC EMISSION TESTING TECHNIQUE

An acoustic emission source generates an expanding spherical wave packet, which is similar to the one that we observe when we throw a small pebble in the calm water of a pond. When these spherical waves reach the body-boundary or the surface of the material under test, a surface wave packet is created as shown in Fig. 7.1.

These surface waves are either Rayleigh or Lamb wave type. A piezo-electric acoustic emission transducer attached to the surface of the body picks up these surface waves. The detection of emitting source, *i.e.,* detection of defect/flaw from anywhere on the surface, allows a higher degree of freedom for detecting the defect from a distance and, therefore, acoustic emission technique becomes very useful for flaw monitoring of flaws which are located at inaccessible places and also for real time monitoring. Other NDT techniques such as pulse-echo ultrasonic technique requires that the probe be right on the top of the internal defect and it may not be possible if the defect happens to be located at an inaccessible place or if the defect happens to be located at a place for which the top surface's contour is either wavy or very complex for the probe to sit over the same. Presence of a defect can be evaluated using a single transducer. However, for linear location we have to go for two transducers and for determination of all the three co-ordinates (x, y and z) for pin pointing the exact location of the defect, one has to use three transducers.

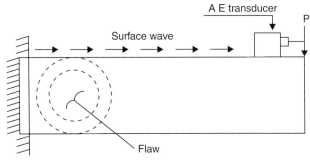

Fig. 7.1

Associated with the acoustic emission testing technique is a parameter called emission count or ring-down count. Acoustic emission count is a weighted measure of acoustic emission activity, which occurs, in a given period of time. A typical method of weighting is to count the number of times the emission signal exceeds a predetermined signal amplitude threshold as shown in Fig. 7.2.

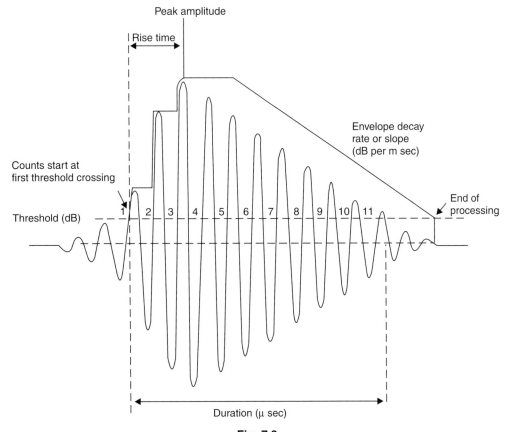

Fig. 7.2

It was referred to as ring-down count because in olden days, a ring used to bell whenever the acoustic signal crossed the predetermined signal amplitude threshold. By this method, a single acoustic emission activity or event is counted several times and large acoustic signals are weighted more heavily than small acoustic signals because large acoustic signal crosses the threshold voltage more number of times as compared to small signal which dies off after crossing the threshold level a couple of times only.

In certain other methods related to acoustic emission testing, instead of measuring acoustic emission counts, events are counted as an alternative. Event is defined as starting at the first threshold crossing and ceasing when there has been no further crossing for a set period of time called the lockout time as shown in Fig. 7.3.

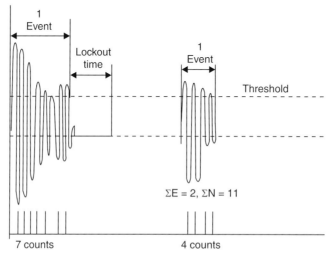

Fig. 7.3

Lockout time is generally very small and is usually of the order of 100 μ sec or so. Both ring-down and event counting may be used with a time-reset system to give an indication of rate of occurrence of emissions. A further simple method is to use a RMS meter to provide mean pulse amplitude as a function of elapsed time during a test.

It is believed that acoustic emission activity (emission rate) and emission energy (emission count) are functions of the deformation rate, absolute stress level and volume of the participating material whereas amplitude of stress waves depends on the energy of the event producing the same.

When a material is loaded in such a way that plastic deformation does take place at the crack tips and other highly stressed regions, acoustic emissions are generated. As is well known, stress concentration factor is very high at and near the tip of any sharp/linear crack and due to this plastic deformation takes place even at loads, which would produce stress in the elastic region only in other portions of the material undergoing loading schedule. This characteristic of acoustic emission technology renders the acoustic emission technique suitable for detection of defects/flaws. Acoustic records containing information about the emission source or for the exact location of the flaw can also be obtained using a number of transducers instead of a single transducer, as has been mentioned earlier. The transducer nearest to the emission source hears the emission first and the time intervals between arrivals of emission signals at various transducers serves as a tool for the exact location of the flaw using software such as triangulation techniques software or alike.

The individual events responsible for acoustic emission for metals are very short lived. These events have rise times in the range of 10^{-4} to 10^{-8} seconds. Hence, an acoustic emission pulse contains a wide range of frequency components extending up to megacycle range. Any transducer, which is sensitive enough to the pressure levels present, will be responsive regardless of its frequency response. Thus, one can choose from a wide band of frequencies and the one most suitable for the test in mind. Most of the acoustic emission work prior to year 1964 was conducted strictly on laboratory basis, with carefully controlled conditions and it was not possible to conduct field tests. The reason for this was that most of the early investigators in the areas of acoustic emission chose to work in the frequency range below 60 kHz (the one which was used by Kaiser) and at this low frequency, it was necessary to have elaborate soundproof chambers, silent loading devices etc. to prevent background noise from interfering with the required signals. Dunegan was the first one to extend experiments into hundred kilocycle and megacycle range to eliminate extraneous noise. Their experiments demonstrated that one could obtain useful result working at higher frequencies. This opened the door for field applications of acoustic emission testing.

The distance between the acoustic emission source and the receiving transducer and the background noise level are primary factors in selecting a particular frequency band. Too high a frequency is not desirable because the high frequency components of the pulse are more severely attenuated than lower frequencies as they travel through the materials. Too low a frequency is also undesirable because background noise then becomes a problem. The optimum frequency band is usually a compromise between these two extremes.

The acoustic emission transducers (or sensor) have piezo-electric crystals. The commonest piezo-electric material used for acoustic emission transducers is Barium-Titanate. However, we do sometime use Lead-Zirconate Titanate (PZT) in place of Barium-Titanate.

7.5 EMPIRICAL RELATIONSHIPS ASSOCIATED WITH ACOUSTIC EMISSION TECHNIQUE

The summation of acoustic counts is known as cumulative acoustic emission counts and is usually written as Σ AE. This cumulative acoustic emission count is a function of applied stress and the relationship between them is as given below:

$$\Sigma \text{ AE} = \text{A}K^{m}$$

where Σ AE = total cumulative acoustic emission during the test,

      A = proportionality constant,

      K = applied stress intensity, and

      m = an empirically derived exponent.

Also, cumulative acoustic emission count (Σ AE), when associated with the formation of a plastic zone at a crack tip, takes the following form:

$$V_{p} = \text{B}\Sigma\text{AE}$$

where V_{p} = volume of the plastically deformed material, and

    B = proportionality constant.

Sometimes, researchers use a parameter 'b' which has a unique value as long as a single type of emission source predominates. If the energies of emission events are randomly distributed, statistical analysis shows that a plot of the number of emissions N_a exceeding a given amplitude level 'a' is generally a smooth curve of the form

$$N_a = (a/a_0)^{-b}$$

where a_0 is a constant and exponential b is a material constant. A logarithmic plot of the cumulative distribution would thus be a straight line of negative slope equal to b.

In the loading of a simple material, a large value of b from the acoustic emission amplitude spectrum indicates emissions from a large number of small events whereas a small value of b shows preponderance of high-energy events. Size effects are eliminated because the value of b is not affected by attenuation of sound in the material, provided that all amplitudes are equally attenuated. In fibre composites, where several types of source events occur, a single value of b is generally not obtained. However, critical points of the amplitude spectrum may be identified by discontinuities in the slope of the b-plot.

7.6 ACOUSTIC EMISSION RESPONSE FROM DUCTILE AND BRITTLE MATERIALS

The acoustic emission response of a brittle material widely differs from that of a ductile material. The acoustic emissions from brittle materials appear in less frequent bursts of higher amplitude than those emitted by a ductile material. The plot showing cumulative counts, as function of loading for ductile materials would show a smooth build up of counts with increasing load. However, the same for brittle materials would show quite periods followed by sudden spurt of emission activities followed by again quite period and so on. As against smooth curve for ductile materials, we obtain step like plot for brittle materials.

During fatigue loading of FRP composites, a good correlation is exhibited between the amount of specimen damage and total acoustic emission. In general, fibre breakage causes emissions of greater magnitude as compared to those generated by either matrix cracking or debonding. However, debonding can sometimes cause relatively large acoustic signal.

7.7 APPLICATIONS OF ACOUSTIC EMISSION TECHNIQUE

Acoustic emission technique may be used to study physical phenomenon such as failure mechanism including crack initiation, crack propagation etc. and also for detecting defects. Hence, this NDT technique is slightly different from other more conventional NDT techniques.

The technique of acoustic emission can successfully be used to locate the source of emission by slightly stressing the component. However, for exact location of the source, a number of transducers have to be used and data thus obtained to be analysed sometimes using 'on-line' computers. Therefore, the technique of acoustic emission for 'flaw' location can be justified only for relatively large and expensive structures where other conventional NDT techniques cannot be usefully employed. The technique has been reported to be in use for testing of large boilers, rocket motor casings, missiles and spacecraft components, pressure vessels and specially FRP pressure vessels. For simple components such as carbon-fibre reinforced plastic laminates or for laboratory specimens, the use of the acoustic emission technique as a flaw detection tool may not be economically viable.

Again, as discussed in the preceding section, acoustic emission response from brittle materials is quite different from that obtained from ductile materials. The components, which become brittle during their life, can thus be checked for their embrittlement and proper heat treatment such as annealing can be provided to restore the component's ductility.

The technique of acoustic emission can be used as an Incipient Failure Detection System (IFDS) as well because as the flaw grows, emissions are released and can easily be detected.

Acoustic Emission can be used for monitoring welding operation both during welding operation and also during the cooling period. The forces causing cracking and hence emissions in these cases are thermal stresses.

Acoustic emission technique can be used for non-destructive testing of adhesive bonds as well. Weaker bonds produce much more emissions at low stress levels compared to the bonds whose surfaces are well prepared.

Acoustic emission has been used widely for studying deformation processes related to fibre-reinforced plastic composites. It has been found that crack propagation rate (i.e., the rate at which plastic deformation takes place) is related to the cumulative count. Generally, fibre breakage causes emissions of higher magnitude as compared with those generated either by matrix cracking or debonding.

Acoustic emission technique is also used for identification and quantification of so called 'latent-defects'.

Acoustic emission technique can be used for monitoring fibre composite pressure vessels as well. A warning about the damaged status of composite pressure vessel is made available in the form of reduced Felicity Ratio (a Felicity Ratio of 0.75 and below is generally not acceptable for fibre composite structures).

The limitations of this technique are that structures which are to be tested under load and the emission depends strongly on the material. Care must also be taken to eliminate spurious sources of emissions. Background noises set a limit to the scale of deformation that can be detected using this technique. Mechanical noise, electrical interference and electronic background noises are also sources of unwanted signals. Though it is easy to record emissions, their interpretation creates problems, because a number of events, which may cause emissions, can occur simultaneously.

7.8 ACOUSTIC EMISSION INSTRUMENTATION/EQUIPMENT

The instrumentation used for acoustic emission studies range from single channel system to the sophisticated multichannel source location system-using computer and visual displays. Block diagram representing the basic units of acoustic emission testing equipment is shown in Fig. 7.4.

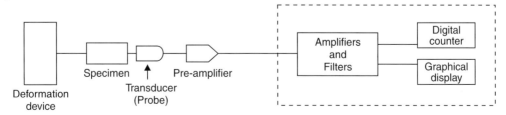

Fig. 7.4

Basically, the detection system has a transducer attached to the specimen. If the emissions from the specimen are of higher magnitude than the prefixed threshold, a transducer picks them out. This received signal is then amplified and read as an acoustic emission count. The response of the system depends upon the amplifier gain setting used and the number as well as amplitude of the pulses being monitored by the transducers. Using a totaliser, the count rate is integrated to give the cumulative emission count ($\sum$ AE).

The acoustic emission transducer converts stress-waves to a low level and high impedance electrical signals. These signals are then routed to an amplifier (total amplification can be up to 100 dB) through a pre-amplifier and appropriate filter. The function of pre-amplifier is of two fold: first to amplify the signal to eliminate environmental disturbances and second to convert the signal into low impedance signal for transmission over long distances. Filters are used to eliminate mechanical (low frequency) and electromagnetic noises. Use of band pass filters allows the choice of an operating frequency range (generally between 100 kHz and 3 MHz) as acoustic emission signals are characterised by their broad bandwidth spectrum. The amplified signal from the amplifier then passes through a threshold detector, which picks only those signals, which are of higher magnitude than a prefixed threshold voltage. Using a totalizer, individual threshold crossings can be summed up to provide cumulative ring down count. One may use strip-chart recorder and oscilloscope also as additions to totaliser unit or as stand-alone system for interpreting acoustic emission data. Sometimes, a digital to analog converter is also used to obtain DC voltage proportional to the number of counts for display on an X-Y plotter. The acoustic emission data can then be plotted directly as a function of other test parameters. Several other accessories such as an audio amplifier, digital printer, analog tape recorder etc. can also be included.

The photograph of the acoustic emission testing equipment is shown in Fig. 7.5.

Fig. 7.5

A frequency window of 750 kHz to 3 MHz is used for acoustic emission transducers in noisy environments. This probe is placed on the specimen and using grease, a proper contact between the probe and specimen surface is obtained. As the specimen is deformed, the transducer picks up emissions and after the signal has been processed, a digital counter gives the measure of cumulative acoustic emission counts.

7.9 ACOUSTO-ULTRASONIC TECHNIQUE

Acousto-ultrasonic technique is a step ahead of acoustic emission technique as far as prediction of ultimate performance specially that of fibre composites is concerned. Acousto-ultrasonic technique is essentially an amalgamation of two separate existing techniques, *viz.*, acoustic emission technique and ultrasonic technique. In acousto-ultrasonic technique, using an ultrasonic pulsar, discrete ultrasonic pulses are injected into the material under test and the ultrasonic pulse is allowed to interact with the material. Due to this interaction of ultrasound with the internal features of the material (or say with the micro-structural environment within the material), the resultant waveform provides nothing but a modulated signal characterising the material quality.

An acoustic emission sensor then picks up this resultant waveform (or the modulated signal of material quality) and treats this modulated ultrasonic signal as if it is an acoustic emission signal where crossing of prefixed threshold amplitude/voltage is counted as a ring-down count. The acoustic emission signal is then suitably processed and digitized in form of a parameter known as stress-wave factor or SWF. This factor provides a measure of material's quality.

Talking in the context of FRP composites, using NDT technique such as eddy current technique, one may evaluate fibre volume fraction in carbon-fibre reinforced plastic composites or using pulse-echo ultrasonic technique, one may locate individual defects in FRP composites. However, it is difficult to isolate their effect on strength characteristics of the material. For example, two same size voids may affect ultimate strength of fibre composite differently, if say fibre volume fraction of two fibre composites sample is not exactly the same. The holistic approach of finding out how defects interact with the micro-structural environment within the material, in which they are present, provides a better way of predicting the ultimate performance of newly made fibre composite. The acousto-ultrasonic technique has got the capability of meeting this requirement.

As described earlier, the ultrasonic input to the fibre composite under test is in the form of well-defined discrete pulses. These ultrasonic pulses are modulated/dampened differently by different fibre composites and also by different features of the same fibre composite. A good quality fibre composite dampens the ultrasonic pulses to a lesser extent as compared to fibre composites having gross distributed defects. The stress-wave factor measurement, which is based on the oscillation counting of the resultant waveform, therefore, is higher for good quality fibre composites as compared to bad quality composite. Defects such as misaligned and broken fibres, micro- and macro-voids, resin crazing, resin rich areas, cracks etc. all reduce the ability of fibre composite to propagate ultrasonic waves and hence, the resultant waveform from a fibre composite containing these defects becomes a highly damped one and the measured stress-wavefactor (which is based on oscillation counting) drops to a lower value for such fibre composites.

Stress-wave factor can easily be used for non-destructive evaluation of ultimate tensile strength of fibre composites. A linear relationship is observed between stress-wave factor and ultimate tensile strength of fibre composites. For fibre composite laminates/sheets, the weakest point on it governs the strength and this is the place where lowest value of stress-wave factor is obtained.

It is also possible to nondestructively evaluate the burst-pressure of fibre composite cylinders using acousto-ultrasonic technique. Experiments have shown that GRP cylinders invariably fail at a place having lowest value of stress-wave factor. Furthermore, a series of tests conducted on a number of GRP cylinders have yielded that a linear relationship exists between stress-wave factor and burst-pressure of GRP cylinders. Hence, one may nondestructively evaluate the burst pressure of GRP cylinders by measuring the lowest value of stress-wave factor for that particular GRP cylinder.

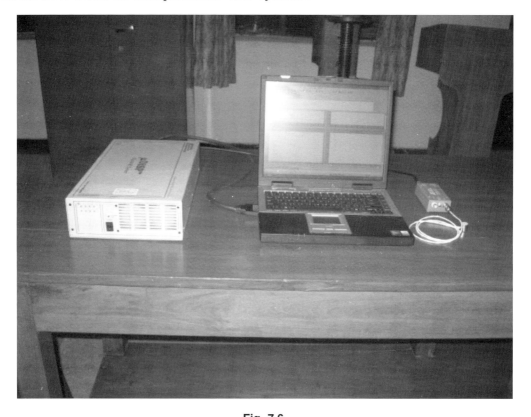

Fig. 7.6

In addition to the above-mentioned properties, certain other properties such as interlaminar shear stress, endurance limit etc. may also be related to stress-wave factor making this acousto-ultrasonic technique a very versatile technique indeed.

The block diagram given shows a typical arrangement of ultrasonic pulsar and acoustic emission sensor along with the related instrumentation for a specimen-undergoing test.

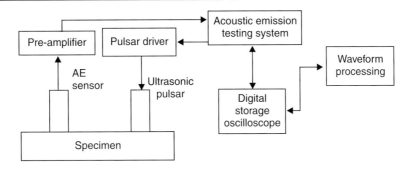

Fig. 7.7 Experimental set-up used in modified acousto ultrasonic technique

Yet another block diagram given below shows the arrangement for the digital readout of SWF parameter, which is a product of parameters g, r and n shown in the Fig. 7.8.

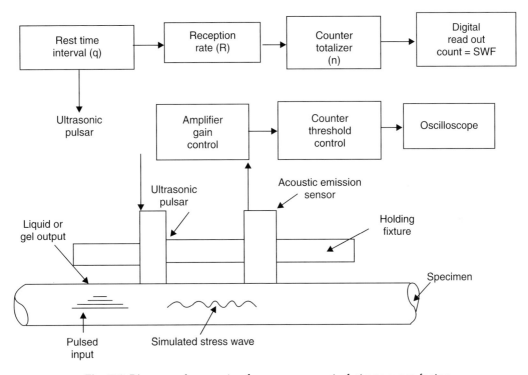

Fig. 7.8 Diagram of apparatus for measurement of stress-wave factor

CHAPTER **8**

Miscellaneous NDT Methods

The NDT techniques which are used very commonly have been described in Chapters 2 to 7. These NDT techniques are called as "Big Brothers" or major NDT techniques because of their wide applicability. Besides these NDT techniques, there are very many NDT techniques which are not commonly used as the ones described in preceding chapters. These so-called miscellaneous NDT techniques or minor NDT techniques, nevertheless, have their own importance. Out of these miscellaneous NDT techniques, only a few NDT techniques have been included in this chapter. These are visual inspection and optical techniques, pressure and leak-testing, resistance strain gauge, brittle coatings, spot testing, spark testing, sulphur printing, thermal methods, electrical methods, dynamic testing, spectro-chemical analysis and Beta ray technique for thickness-measurement.

8.1 VISUAL INSPECTION AND OPTICAL TECHNIQUES

Visual inspection technique is one of the oldest and simplest NDT technique. It is simple in use, easy to apply, quick in interpretation and a low cost NDT technique. It is very commonly used for the very first inspection of any object which comes for non-destructive evaluation.

In its simplest form, all that one needs is his own eyes. Visual inspection is mostly used for ascertaining the type and location of defects, if any, present at the surface of the object. Besides its use for detection of surface flaws, the technique is also used for checking surface finish, scratches, decolouration, through surface openings, finish of papers and objects alike, assessment of internal defects in transparent materials (*e.g.,* inspection of hand laid up low volume fraction glass fibre reinforced epoxy laminates), strained and crazed areas in transparent polymeric materials etc. Sometimes, visual examination precedes other NDT inspection and is of great help (*e.g.,* visual examination of weld bead aids in deciding the angle of incidence of X-ray beam while examining the weld for internal cracks).

Cleaning of the surface of the object which is to undergo visual inspection, is a must for proper inspection. One may sometime require operations such as sand blasting or shotblasting for cleaning the surfaces. If the surfaces are not cleaned properly before undertaking the inspection, certain surface defects may remain undetected.

The inspection area for visual method of NDT should be properly illuminated for quick inspection. One may employ drop lights for proper illumination of the required area. Also, the period of time during which a human inspector is to inspect, should be limited to minimise the human visual errors. The minimum size of defect, that can be detected by naked eyes, depends upon the surface being inspected, the level of brightness and the brightness-contrast between the specimen and its background.

While adopting visual inspection NDT technique, one must remember that normal eyes cannot focus sharply on objects which are closer than approximately 0.1 mm. Otherwise, nearness of the object provides larger visual angle and therefore better visual inspection. Also, the minimum angular separation of two point objects resolvable by the eye, without any optical aid, is about one minute of arc.

For enhancing the capabilities of visual inspection technique, many optical aids are used. These optical aids are mirrors, magnifying lenses, microscopes, periscopes, telescopes, boroscopes, endoscopes etc. These optical aids or techniques are employed for enlarging the tiny surface defects and thus aid in easy viewing.

Sometimes, light sensitive devices such as photocells and phototubes are used for replacing direct visual inspection and thus compensate for errors due to operator's fatigue.

Mirrors are used for looking into inaccessible and forbidden area. While using mirrors, it must be ensured that the surface of mirror is extremely flat and free from dust particles to avoid spurious indications.

Magnifying lenses are used to magnify surface features. Magnifying lens should be held as close to the eye as possible. This closeness permits the greatest number of rays from the specimen to enter the eye and minimise the glare. Magnifying lenses are different than reading lenses which are not kept close to the eyes.

Microscopes are used to obtain greatly magnified image of small objects. For even higher magnifications, electron microscopes are employed. Periscopes are instruments in which the general direction of the rays is not a straight line but is deflected, one or more times, for the purpose of providing the observer with a view from a position in which he cannot put his head.

Telescopes are used to magnify image of a distant object or to gather more light than that reaches the unaided eye. Boroscopes permit direct visual inspection of the interior of hollow tubes, chambers and other internal surfaces. Boroscopes are precision built optical systems having a complex arrangement of prisms, achromatic and plain lenses.

Endoscopes are typically 5 to 10 mm in diameter and 150 mm to 1 m in length. Endoscopes are rigid viewing devices and they generally have interchangeable lenses at the viewing end, to look forward, sideways, backward and so on. While using endoscopes, one may view through naked eyes or with the help of still, movie or video cameras. The light from a small optical source near the eye-piece can be carried down to a projection lens at the tip by a rigid plastic, glass or quartz light guide or via a bundle of optical fibres and viewing directions may be controlled from the eye-piece.

Flexible viewing devices employ optical fibre bundle, both to transport the incident light and to view the reflected image. In fibre optics devices, however, the resolution of the image cannot be better than that provided by the diameter of individual optical fibre which is of the order of 0.01 mm. Also, due to absorption of light by the glass; total length of fibre optics devices is generally restricted to about 3 m or so. Finally, the very versatility of the fibre optics devices in reaching inaccessible locations means that it becomes very difficult to know where the end is and in which direction is one looking? It becomes difficult to differentiate between a pit and a blister, while using any remote viewing aid.

8.2 PRESSURE AND LEAK TESTING

The most common example of pressure and leak testing which one observes very frequently is while getting his bicycle, scooter or car puncture mended. The tube is first inflated by applying proper air pressure (*i.e.,* higher than external pressure) and then by dipping the inflated tube in water-tank, the leak is located by searching for the presence of air-bubbles. This common day practice is nothing but a very simple type of pressure and leak testing technique.

In general, in pressure and leak testing, leaks are detected by following basic methods:

(*i*) Looking for water or gas seepage.

(*ii*) Observing the changes in pressure of the liquid or gas being used for pressurisation.

(*iii*) Searching for bubble formation when the test object is covered with soap solution or immersed in liquid.

There are several type of pressure and leak testing techniques such as air pressure testing, hydrostatic testing, radioactive leak testing, penetrant leak testing, helium leak detection technique, sonic and ultrasonic leak detection technique etc. Materials Evaluation (Published by The American Society for Non-Destructive Testing), Vol. 43, No. 2 February 1985 describes some of latest commercially available leak testing instruments under "Product Showcase" and these leak testing instruments are based on the principles of one of the aforementioned techniques.

In air-pressure testing, air is introduced into the vessel under test and leaks are observed by means of soap solution applied outside the vessel. Sometimes, instead of applying soap solution, the vessel is immersed under water. The sensitivity of the test increases at higher pressure. When water is used for immersing the vessel, larger bubbles are formed and they take so long to appear at the surface that they may easily be missed. Therefore for immersing the vessel, liquids having low surface tension should be preferred. In low surface tension liquids such as ether, alcohol, acetone, methanol, soap-solution etc., the bubbles which form are about five to six times smaller in diameter and therefore, these bubbles are emitted at very rapid rate. As these bubbles move slowly and form at a rapid rate, they appear as a vertical stream which is easily noticed.

For air pressure testing, generally compressed air-lines are not recommended because dirt and oil particles present in it may block small leaks temporarily and they may remain undetected. Soap solutions may also pose a similar problem. Finally, it is recommended that the air pressure be applied before immersion so that the fluid does not enter small leaks beforehand.

In hydrostatic testing, which is the simplest and most commonly used pressure and leak testing method, water or water containing dye is used for applying gradual hydrostatic pressure. The amount of pressure to be applied is generally governed by different codes and specifications and is generally 1.5 to 2 times working pressure of the vessel. By this method of testing, generally only the large defects such as centre line cracks in welds and pin holes are revealed. Fine cracks may remain undetected by this technique.

In radioactive leak technique, the vessel or pipe like specimens to be tested are placed in a tank and the tank is sealed thereafter. Through the inlet of this sealed tank, a radioactive gas is allowed inside the tank at certain recommended pressure. Sufficient time is allowed for radioactive gas to penetrate through the leaks or holes into the specimens. Finally the specimens are removed from the tank and their surfaces are cleaned to remove the traces of radioactive material at the surfaces. This step is similar to the removal of excess penetrant from the surface, as was described under liquid penetrant inspection (Chapter 5). Using a radiation detector, one may then detect any accumulation of radioactive gas within the specimen and thereby detect the leak.

In another radioactive leak technique, which is used for detecting leaks in pipelines; radioactive liquid is pumped into the pipeline from one of its end. Thereafter, some time is allowed for the pumped liquid to flow a distance of about 2 km or so. A battery operated portable radiation detector cum recorder is then put in the pipeline. This radiation detector is carried from one end of the pipeline to the other end by the flowing liquid. This detector cum recorder enables the operator to locate the position(s) where the radioactive solution has leaked. While using radioactive leak technique, precautions to be observed for handling radioactive materials should strictly be followed.

Dye and fluorescent penetrants which have already been described in Chapter 5, may also be employed for the detection of through leaks in tubings and similar type of vessels. This type of leak testing is called penetrant leak technique.

Helium gas is very successfully used to detect very minute leaks and the instrument used for this purpose is known as helium leak detector. Helium leak detectors are basically portable mass spectrometers which are highly sensitive to the presence of helium gas. These leak detectors can detect the presence of less than one part of helium in ten million parts of air. Helium being a light gas, passes through small leaks more easily than the heavier gasses. As helium is not present in any significant quantity in the atmosphere, working of helium leak detectors remain unaffected by the surrounding air. Finally, helium being an inert gas, it does not react with the gases which may be present in the vessel to be tested or does not affect the material of the vessel. In helium leak detection technique, one may either adopt pressurisation technique with helium as pressurisation fluid or one may adopt the accumulation technique with helium being used in place of radioactive solution.

In sonic leak detection technique, the principle involves fully pressurising the vessel under test and detecting the sound produced by leaking gases by unaided ear (less sensitive) or by suitable electronic devices (more sensitive). Sonic leak detection technique has one draw back that the background noise limits its capabilities.

In ultrasonic leak testing, the principle involved is the same as that for sonic leak testing. However, the working frequency of the ultrasonic leak detectors is of the order of 80 kHz which is well above the sonic (audio) range of 30 Hz to 20 kHz. The broadband noise is produced due to turbulence of gases leaking through tiny holes. As the instrument works in the ultrasonic frequency range, mechanical and air-borne noises do not affect the detection. The ultrasonic leak detectors are typically of the size and shape of a torch. This technique is a non-contact type and is recommended for tiny leaks only. For leaks, soap solution technique described earlier provides better and quicker results.

8.3 RESISTANCE STRAIN GAUGE

Resistance strain gauge technique is generally looked at as a stress analysis technique, rather than a non-destructive testing technique. This is because of the fact that normally a non-destructive test engineer is more interested in locating external and internal flaws or in some form of non-destructive evaluation of physical properties etc. (*e.g.,* thickness measurement from one side, measurement of fibre volume fraction and lay-up order determination in fibre composites etc.). In the present day's concept, the job of strain measurement using strain gauges, for the purpose of stress analysis belongs to stress analyst rather than to a non-destructive testing engineer. However, one would certainly agree to the fact that as per the definition of NDT techniques, any technique in which the future usefulness of the material or the product under test is not impaired, should come under the purview of non-destructive testing techniques. Since the use of strain gauges does not impair the future usefulness of a product or component, resistance strain gauge technique ought to be considered as a non-destructive testing technique. Furthermore, one should bear in mind that NDT techniques are not only for the defect analysis but are also for evaluating different parameters non-destructively. Since, electrical resistance strain gauges evaluate strain values nondestructively, they very much become part and parcel of NDT techniques.

In applications where theoretical stress analysis of complex shape structure, or structures made of anisotropic materials becomes practically impossible, experimental stress analysis paves the way for evaluating the stresses. Also, the experimental techniques (such as resistance strain gauge technique) are being extensively used for verification and measurement of stresses and strains and they have become very vital for the modern technology in all fields of engineering applications.

Resistance strain gauge technique is by far the most popular strain measuring technique today. Besides being used for measuring strains, the resistance strain gauges are also used for evaluating parameters such as force, torque, pressure, speed, acceleration, surface finish, straightness, displacement, dimensions, blood pressure etc. Resistance strain gauges can be used on almost all metals, ceramics (*e.g.*, cement, concrete, bricks, glass etc.), polymers, ribbons, wood, bone (live and dead) etc. They have been used for stress-analysis/strain monitoring purposes on aeroplanes, ships, submarines, big reciprocating engines, gas turbines, automobiles etc.

Resistance strain gauges are basically hair thin wires of nickel based alloys (*e.g.*, nickel chrome, medium nickel chrome, nickel iron, copper nickel etc.) The fine wire used for strain gauge applications is not in the form of a straight wire but is arranged in various patterns such as flat grid, wrap around or bobbin type, dual element, rosettes etc. This very thin wire, arranged in a particular pattern, is bonded to paper base on its top and bottom. Outcomes, from these paper backings, the two ends of the wire, for subsequent electric connections.

While using strain gauges, the gauge is firmly fixed to the component under test, at a place where strain values are desired, using epoxy resin (araldite) as adhesive. The adhesive bond should be without any air bubbles, rough spots or inclusion. The surface to which the gauge is to be fixed must be smooth and even but should not be a highly polished surface. Dirt, scales, rust, grease etc. should be removed from the surface and the surface should then be cleaned using acetone or carbon tetrachloride soaked clean cotton piece. One should keep changing these cotton pieces till these soaked cotton pieces do not show any trace of dirt on them. One must remember that for perfect adhesion, it is absolutely essential to have extreme cleanliness.

After fixing the strain gauge to the desired location, electric current is passed through the gauge. The resistance of the gauge varies with the applied stress. This change in resistance of the gauge is correlated with the stress produced within the structure under test. One may either record the change in resistance or directly read the strain readings. Potentiometer,

Wheatstone bridge, a recorder, an oscilloscope or a digital strain indicator is used for this purpose. It is of utmost importance to use a sensitive circuitry and/or a reliable measuring instrument to obtain correct strain values.

Details regarding Wheatstone bridge circuitry and their balancing arrangement of strain gauges for different loading situations, and for maximum circuit sensitivity, temperature compensation etc. can be obtained from any standard textbook on resistance strain gauges. There are a number of books on strain gauges and a few of them are listed below for ready reference to strain gauge technique.

1. Dally, J.W. and Riley, W.F., *Experimental Stress Analysis*, New York, McGraw Hill, Inc., 1991.

2. Hetenyi, M. (Ed.), *Handbook of Experimental Stress Analysis*, New York: John Wiley and Sons, Inc., 1950.

3. Lee,G.H., *An Introduction to Experimental Stress Analysis*, New York: John Wiley and Sons, Inc., 1950.

4. Perry, C.C. and Lissner, H.R., *The Strain Gauge Primer*, New York: McGraw Hill Book Co., Inc., 1955.

5. Window, A.L., *Strain Gauge Technology,* Barking, Essex, Elsevier Science Publishers Ltd., 1992.

8.4 BRITTLE COATINGS

Brittle coatings (also known as brittle lacquers) can be imagined as a sort of thin and quickly drying paint which can easily be sprayed. These coatings fail at a particular value of strain (known as threshold strain) and each brittle coating has its own characteristic threshold strain value. Hence, if a component, which has been designed to withstand a strain value of ε, is sprayed with a brittle coating having threshold value lower than ε, the coating shall fail before the component reaches its failure strain. This way, the brittle coating can act as a non-destructive surveillance system. In the same manner it can also detect overloop regions non-destructively. Furthermore, since the spraying does not impair the future usefulness of the component, the brittle coating technique becomes one of the non-destructive testing techniques.

Besides being used for providing warning about the impending failure or overstraining, brittle coatings are extensively used for obtaining graphical representation of the distribution, direction, location and magnitude of tensile strain, over the entire surface of a two-dimensional or three-dimensional component.

The brittle coating method on non-destructive testing is applicable to all types of metals and metallic surfaces, glass, wood and plastics. The coatings crack normal to the principle tensile strain direction and the first crack appears when the strain in the component reaches the threshold strain value of the coating.

For obtaining the value of threshold strain of a particular brittle coating, a test bar is first sprayed with that particular brittle coating and after the coating has dried-up, the bar is deflected by a fixed amount in a commercially available cantilever-beam test fixture. The test bar is then placed in a strain scale and the minimum amount of strain required to fracture the coating is noted. This strain value is called the threshold strain of the coating.

There are two types of brittle coatings, *viz.,* resin type and ceramic type. Resin type brittle coatings are very widely used in the moderate temperature range whereas for high temperature applications, ceramic type brittle coatings are used. As far as accuracy of the test is concerned, resin type coatings provide for better accuracy of strain measurement. Additional quantitative accuracy can be achieved by subsequent usage of resistance gauges, described in the earlier section.

The brittleness of brittle coatings is effected by both the temperature and humidity conditions during spraying, drying and testing. Therefore, it becomes necessary to select and calibrate proper coating for every test. For selecting purpose, the manufacturers of brittle coatings provide suitable charts using which one can select the proper brittle coating for the conditions prevailing at the time and place of testing. The commercial coatings usually come in aerosol cans and can easily be sprayed onto the components surface. The advantage of spraying, as against brush painting, is that a uniform thickness of the coating is obtained.

Before spraying the coating, surface cleaning is required, as in case of most of the other non-destructive testing techniques. Dust particles, loose scales, grease and oil must be removed from the surface where brittle coating is to be sprayed. The cleaning operation is essential for proper evaluation of strain field.

Spraying should be done at temperatures equal to or higher than those expected during the test schedule. Sometimes, the components to be tested are heated up to about 50°C before the brittle coatings are sprayed. This is done to minimise the problems associated with humid weather. After spraying, the coatings are allowed to dry at a temperature which is somewhat higher than the test temperature. The drying operation takes about 12 to 24 hours. While spraying the test component, calibration bars are also sprayed in the same conditions of temperature and humidity as the component. These sprayed calibration bars should be dried and tested alongwith the component. When the drying operation is over, the component is put to the test and as soon as the component develops a strain value equals to the threshold strain level of the brittle coating sprayed, cracks develop in the coating. These cracks generally remain opened once they are formed and can easily be viewed under proper illumination. The illumination should preferably come from a point approximately tangent to the surface under observation. Flashing lights make the cracks to shine.

High threshold strain brittle coatings produce very tightly closed cracks (after the removal of load) and their viewing becomes practically impossible with the naked eyes and illumination arrangement. For such applications, dye-etching technique or electrified particle technique is required to be used for easy viewing of the crack-pattern.

If one wants to have a permanent photographic record of the crack pattern, use of crayons or dye-pen is made. One may also mark the values of strain/stress at which these patterns have developed and these data may be written on the component itself. These markings with crayons or dye-pens enable one to obtain the complete test data in the same photographic record.

8.5 SPOT TEST

Spot test technique is one of the chemical methods of non-destructive testing. It is also considered as one of the rapid identification techniques for the classification of metals and alloys. Rapid identification techniques play very important role when one is required to quickly sort out different types of metals and alloys from a heap of scrap metals or when scrap metals/ alloys are to be salvaged. Spot tests do not provide a quantitative analysis of various constituents in an alloy (*i.e.*, the alloy composition). Spot test technique is only a qualitative technique for grouping/identification purpose.

Spot test technique is applicable to both inorganic and organic compounds. Fritz Feigel has published two volumes on Spot Tests. The book is entitled "Spot Tests" and has been published by Elsevier Publishing Company, Amsterdam, 1954. The first volume deals with inorganic applications whereas the second volume deals with organic applications.

Spot test technique is based upon the chemical reactions between a test solution and a reagent. For identifying different metals and alloys, different reagents are used. Feigel's book provides the details about the reagents to be used for different metals and alloys.

Generally, in spot test technique, a drop of recommended reagent solution is put on a small quantity of solid specimen and the reaction which takes place provides a clue for the identification of the material. Sometimes, instead of using a drop of reagent, a strip of reagent paper is used. This reagent paper is exposed to the gases which come out from the test solution or from minute quantity of solid specimens. The change in colour of the reagent paper then provides the clue for the identification of metals and alloys.

Spot test techniques has their own limitations and cannot act as a substitute for finer techniques such as eddy-current technique for material sorting, spectrochemical analysis etc. Whenever accuracy of identification is required, spot tests cannot be recommended. As mentioned earlier, spot tests should be used only when rapidness rather than accuracy is the criterion (*e.g.*, salvaging of scrap-metals).

8.6 SPARK TESTING

This is one of the very old method for the classification of different types of steel. Like spot test, spark testing technique comes in the category of rapid methods for identification of metals and alloys. Despite being an old technique, spark testing technique still does find wide applications for classification of steels in various alloy steel/steel plants.

As the name suggests, spark testing technique is based upon the sparks produced by the test specimens. When steel specimens are held against an abrasive wheel revolving at high angular speed, sparks are produced. The spark emission phenomenon is observed very commonly when one is getting the blades of kitchen knife sharpened. Similar spark emission phenomenon is observed while watching surface grinding operation on steel plates etc. or while steel is cut using abrasive wheel cutters. These sparks are produced by the metallic particles being removed by the abrasive wheel from the bulk material. The removal of metallic particles takes place by tearing process and the temperature of the metallic particles get raised to incandescence (glowing) due to very rapid rate of tearing.

Different steels produce their own characteristic spark (the spark pattern and the colour). One may observe the spark characteristic a number of times to get accustomed to the spark pattern of known steel samples and later on use his judgment for rapid identification of different steels. Otherwise, one may observe the spark pattern and compare the same with reference photographs/sketches of the spark characteristic of different grades of steels. Metals Handbook published by the American Society of Metals also contains these reference photographs/sketches for different steels. One can produce his own set of reference colour photographs by taking pictures of the spark emission pattern for different steels samples being commonly produced or used in his plant/workshop and later use this set of reference photographs for comparing the spark emission patterns of unknown steel samples. This shall help in rapid identification/ classification of different types of steel.

As viewing of sparks produced is done by the inspector or NDT engineer, human factor is involved in spark testing, *i.e.,* wrong viewing or error in judgment may lead to error in identification or classification. This will result in mix-ups and may lead to further trouble. Therefore, spark testing should be performed by a person with good eye-sight. An experienced person only can perform spark test quickly and with certainty. Such persons need only a very quick look at the instantaneous spark emission to rapidly classify different steels. For more accurate analysis of different types of steels, one may opt for other techniques such as eddy-current testing, spectrochemical analysis etc. Otherwise, as far as rapidness of classification is concerned, spark testing is most suited and is, therefore, very widely used even today in many steel plants/industries.

8.7 SULPHUR PRINTING

Sulphur printing technique is one of the rapid identification techniques for the classification of steels and different types of alloy steels. As is suggested by the name of the technique, it is based upon the presence and distribution of "sulphur" in the samples. The process involves obtaining "prints" of sulphur rich areas on silver bromide photographic printing paper. The nomenclature also suggests that just as finger-prints are used for identification purposes, sulphur-prints are used for identification of the type of steel.

As was mentioned in the case of spark testing, for sulphur printing technique too, one may obtain a set of reference sulphur prints (photographs) for a number of commonly used and known steel samples. Thereafter, one may use this set of reference sulphur prints for identifying purposes by way of comparing the sulphur prints of unknown steel samples with the set of standard sulphur prints.

In sulphur printing technique, first of all, a 3 to 4 per cent sulphuric acid (H_2SO_4) solution is prepared. While preparing this acid solution, one must remember that concentrated sulphuric acid causes severe burns if handled carelessly. Therefore one should handle sulphuric acid with utmost care. While making 3 to 4 per cent acid solution, it is important to add acid to water and not vice versa because in the later case danger due to spattering is always present. For dilution purpose, either a glass container or a porcelain pot should be used because of the highly corrosive nature of sulphuric acid.

Once the acid solution is ready, a sheet of silver bromide printing paper is thoroughly wetted in this acid solution. This wet (moist) photographic paper is then held in an upright position for draining-off the excess acid solution. The steel samples, which are to be tested, are then placed over the moist printing paper (on the emulsion side). The samples are allowed to lie there for about 5 minutes. This time is allowed for sulphuric acid to combine with the sulphur present in the steel samples to produce hydrogen sulphide which in turn reacts with silver bromide to produce silver sulphide which is brown in colour. Thus, after the samples are removed, one observes brown colour patterns on the photographic paper. The intensity of brown colour depends upon the percentage of sulphur present in the steel samples. As mentioned earlier, using known steel samples, one may obtain a set of standard sulphur prints for reference purposes. Later on, one may use these reference sulphur prints for identification of unknown steel samples, by way of comparison.

The surface of steel sample to be tested should neither be a very polished surface not a very rough surface. Ordinarily, roughly polished surface are used for sulphur printing tests. Also, the contact surface should be free from dust, rust, scales and grease etc.

Once the sulphur testing is over, samples should be thoroughly washed, preferably under hot running water to remove any acid remaining over the surface. This is required to avoid any subsequent corrosion of the samples tested. Scrubbing with a wire brush in addition to washing is a better way of ensuring thorough cleaning. As a further protection to the tested steel samples, one may even wash the tested surfaces with dilute ammonia solution to neutralize any acid remains and thereafter wash the samples thoroughly using the wire brush and hot running water. After washing operation, the samples should be dried properly and then a thin clean lacquer such as vaseline or good quality grease etc. should be applied all over to provide protection against atmospheric corrosion (rusting).

The technique of sulphur printing in which sulphuric acid solution and photographic paper are used is also known as Baremann sulphur printing technique. However, the nomenclature "sulphur printing" is more commonly used and is better understood.

In another sulphur printing technique known as Heyn and Bauer sulphur printing technique, instead of using sulphuric acid solution, an acid solution of mercuric chloride is used. Also, instead of wetting a photographic paper, an ordinary silk rag is wetted. The silk cloth soaked in mercuric chloride solution is then pressed on the steel specimens. The surface preparation requirements are the same as described earlier. The silk cloth remains pressed onto the steel specimens for about 5 minutes. Once the silk cloth is removed, one observes black colouration in certain areas on the surface of steel samples. These black areas refer to sulphur rich areas. In this technique, besides sulphur rich areas, even the phosphorous rich areas are identified because the phosphorous rich areas turn yellow in colour as the sulphur rich areas turn black in colour.

Baremann technique has two advantages over Heyn and Bauer technique. The first advantage of Baremann technique is that a permanent record of the test conducted is obtained and the second advantage is better contrast and, therefore, easier viewing and subsequent classification.

Finally, as was in the cases of spot testing and spark testing, only qualitative results should be expected out of sulphur printing technique.

8.8 THERMAL METHODS

Thermal methods of non-destructive testing are based on the principles of thermal sciences and basically in all the thermal techniques of non-destructive testing, the specimen to be tested is suitably heated and the resulting temperature distribution is studied for finding flaws etc. Flaws such as porosities (blow-holes or voids), internal cracks etc. are nothing but entrapped air within the specimen and thermal conductivity of air and metallic specimens are widely different. Therefore, presence of flaws results in abrupt change in temperature distribution. Thermal methods are applicable not only to metallic materials but to polymers and ceramics too.

For heating the specimens under inspection, one may opt either for conventional heat sources or go for infra-red heating devices, induction heating, electrical heating or alike. For studying the temperature distribution, one may use thermocouples, temperature sensitive phosphors and paints, wax, infra-red films, resistance thermometers, photoconductive materials etc.

One of the thermal methods of non-destructive testing employing temperature sensitive phosphors such as zinc cadmium sulphide phosphor is known as "thermography" which is nothing but temperature sensitive phosphor technique. In this technique, a very thin layer of temperature sensitive phosphor paint is sprayed over the test samples, just like spraying paint on car bodies. This is important to make sure that the layer is thin. As the paints are bad conductor of heat, thicker the layer, less sensitive shall be the detecting method. Also, one should remember that the colour-change process in phosphor coatings is a reversible process and therefore one should be very alert while studying the temperature distribution or temperature changes. Due to this reversible nature, no permanent record can be maintained either. However, if permanent record is definitely required, photographic techniques may be put to use. Generally, while using phosphors, sensitivity of the testing technique should be checked using samples having known defects.

Besides using temperature sensitive phosphors, temperature sensitive crayons with characteristic melting points, are also used for studying the temperature distribution. Such crayons are commercially available for different temperature indications. The accuracy of temperature indication by such crayons is within plus and minus 1% of its rated temperature. In this technique, the sample to be tested is first marked with suitably rated crayon and then the sample is heated. As soon as the temperature of the sample reaches the rated temperature of the crayon used, the crayon melts. It is the melting phenomenon that is the indication to be observed and not the indications such as change of colour etc. In the areas containing flaws, melting does not take place at the same instant and thus the flaw can be located.

Sometimes, instead of using commercially available temperature sensitive crayons, temperature indicating lacquers are also used. These lacquers may be imagined as temperature sensitive crayon particles suspended in a volatile carrier fluid, which enables one to spray it on the surface of the sample to be tested. Because of the volatile nature of the carrier fluid, the temperature sensitive lacquer dries off very quickly and it leaves a dull looking thin opaque coating at the surface. This situation is similar to the one in which temperature sensitive crayons are applied to the surface.

Temperature sensitive pigments which change their colours with temperature, are also used as temperature indicating devices for studying temperature distribution, while employing thermal methods of non-destructive testing. These temperature sensitive pigments, also known as thermal colours, are commercially available and they possess the characteristics of changing their original colour with the rise in temperature. This change in colour is not something like from red to light red or dark red or dark pink (i.e., not different shades or the same colour) but something like blue to yellow, yellow to black, black to green etc. (i.e., distinct changes in colour). For example, one commercially available thermal colour which is originally pink coloured becomes light blue at 65°C, yellow at 145°C, black at 175°C and green at 340°C. These changes in colour are permanent and can thus provide a record of the test conducted. However, after the test schedule is over, rapid cooling is recommended, if a permanent record of the test is desired.

Infrared thermal method, also known as infrared thermography is one of the most popular thermal methods of non-destructive testing. As is well known, the infrared region is beyond the visible spectrum and infrared literally means below red. The lower wavelength infrared radiation (*i.e.,* the ones, near to visible spectrum) are generally used for infrared thermography and this infrared radiation can easily be detected using photographic means. Special infrared cameras and infrared photographic films are commercially available for this purpose. Infrared photography finds wide application in military and geological surveillance too because infrared photography depends upon the reflection of infrared rays from the surface of the object to be photographed. Also, as human skin is transparent to infrared radiation, infrared radiation devices are used even for diagnosing abnormal conditions immediately below the skin-surface.

In infrared thermography, an infrared camera having infrared colour film is used to study the temperature distribution. On developing the film, different colours indicate different temperature zones and thus one can locate the areas containing defects etc.

8.9 ELECTRICAL METHODS

The electrical methods of non-destructive testing are based upon the basic fact that the electrical resistivity of a material changes considerably in the area where the flaws are situated.

It is well known that when an electric current flows through an electrically conducting material, it results in potential difference between any two points on the material. The presence of a flaw or flaws change(s) the magnitude of this potential difference. For the electric current to pass through one point to another point special electrodes are required. These electrodes are so designed that the surface contact resistance is reduced to a non-significant value. If these special probes are not used, the variations in the surface contact resistance overshadow the variations in electrical resistance due to the presence of flaws and thereby making it practically impossible to evaluate the presence of flaws.

The following well known expression relates the potential difference V (volts), current I (amperes) and resistance R (ohms):

$$R = \frac{V}{I}$$

Also, there is another expression relating V, I and resistivity ρ (ohm-metre):

$$\frac{V}{I} = K \rho$$

where K is a constant of proportionality and has unit of length (centimetre) and this constant depends only upon the geometry of the specimen and the geometry of the electrode arrangement.

For a fixed electrode arrangement and for the same material (*i.e.,* ρ is constant), any change in the ratio of V/I will, therefore, represent a change in the geometry of the part. This is the basic principle involved in electrical method of non-destructive evaluation of wall thickness from one side only. This principle is used for evaluating thickness of plates, boiler tubes, pressure vessels etc. having only one side accessible.

In other electrical methods of non-destructive testing, potential difference is measured as a non-destructive test parameter. Potential difference technique is used to evaluate the homogeneity and soundness of bars, tubes, wires, large turbo-generators, railway tracks, railway axles, castings etc. By measuring potential difference across welds, welds can be inspected for detecting cracks, improper weld penetration and lack of adhesion of the weld metal. In such testings, the potential difference across defective welds is far greater than that across a good weld.

If the dimensions of the sample under test are uniform, the constant of proportionality K in the previous equation can be replaced by L/A where L is the length of specimen (metre) and A is the area of the cross-section of the sample (square metre). Thus, the previous equation can be rewritten as

$$\frac{V}{I} = \frac{L}{A}\rho$$

or,
$$\frac{V}{I} = R$$

For finding the magnitude of an unknown resistance, generally Wheatstone bridge is employed. A schematic representation of Wheatstone bridge is shown in Fig. 8.1 (a). In this schematic representation, E represents a battery which provides electric current. This current is sent through unknown resistance R_X and then through the known or standard resistance R_S.

The battery also sends current through resistances R_1 and R_2 as shown in the same figure. A galvanometer (G) is connected between points B and C to observe the balance. When using Wheatstone bridge for evaluating the unknown resistance (R_X), first of all resistances R_2 and R_1 of suitable values are selected. Thereafter, R_S is adjusted till a balance is achieved, *i.e.,* galvanometer shows a zero deflection. In this balanced position, the following relationship holds good.

$$R_X = R_S \frac{R_1}{R_2}$$

In another type of Wheatstone bridge, which is known as slide wire Wheatstone bridge [Fig. 8.1 (b)], the resistances R_1 and R_2 are replaced by a high resistance wire. A sliding contact joins the galvanometer to the movable point P. As the wire used has uniform cross-section, the resistances of segments M and N become proportional to their lengths and, therefore, the previous equation gets modified as

$$R_X = R_S \frac{M}{N}.$$

While employing Wheatstone bridge, one should keep in mind that the current through unknown resistance should be as large as the resistance will permit without undue heating. Wheatstone bridge is generally used for measuring resistance having magnitude greater than 1 ohm. For measuring smaller magnitude resistances, Wheatstone bridge technique is unsuitable.

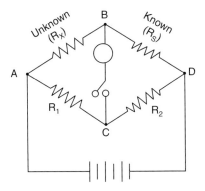

Fig. 8.1 (*a*) Schematic representation of a Wheatstone bridge

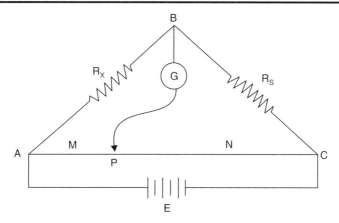

Fig. 8.1 (*b*) Slide wire Wheatstone bridge

Electrical methods of non-destructive testing may be used for evaluating the constituents of metallic alloys also. This is based on the fact that the conductivity of a pure metal decreases appreciably when the alloying elements are added to it. In these methods, basically the conductivity measurements are involved and are calibrated against the alloy constituents such as estimation of magnesium in aluminium alloys or amount of silicon in silicon iron sheets, etc.

Electrical methods can also be used for monitoring corrosion rate. The procedure basically involves having an electrode in contact with the surface whose corrosion is to be monitored. As corrosion removes metal from the surface of the specimen, the electrical resistance (contact resistance) increases and this increase in electrical resistance can be calibrated against corrosion rate.

8.10 DYNAMIC TESTING

There are many test methods which come under the category of dynamic testing methods of non-destructive testing. One of the very old dynamic non-destructive test methods, which consists of listening to the sound produced by hammer or mallet tapping on the specimen, is known as mallet tapping or coin tapping technique. The sound produced by so called "mallet tapping" is quite different in case of a "defect free" specimen as compared to specimens having defects. This technique is, however, very crude and may involve human errors. Hence, mallet tapping technique is not considered to be a reliable technique.

To eliminate the human errors, one may opt for electronic instruments to listen to sound pattern produced by mallet tapping. These electronic instruments shall basically be frequency measuring instruments or frequency deviation meters. A defective sample shall provide a flat frequency response as compared to a defect-free sample of the same material. However, the technique still does have the drawbacks such as the frequency of emitted sound depends upon the way the specimen is supported, the way the specimen is tapped and the amount of striking force etc. With so many variables, standardisation of the technique becomes a problem.

In second type of dynamic testing, instead of evaluating the frequency response, the attenuation rate of vibrations in solids is used as the measuring parameter. For initiating the vibrations, one may adopt simple impact technique or may employ better ways such as use of piezo-electric transducers, electromagnetic vibrators and induced current vibrators. In electromagnetic vibrators and induced current vibrators, the specimen is free to vibrate on its own because no physical contact is required between the specimen and the exciting system.

Once the specimen starts vibrating, its free oscillations decay even when it is isolated from the surrounding environment. This process is referred to as damping phenomenon. The damping capacity of a specimen depends upon many factors such as magnitude of applied force, frequency of vibration, surrounding temperature, metallurgical conditions such as composition, grain size, heat-treatment etc. Generally, the damping capacity of an alloy is less than its constituent elements. For solid solutions, the damping capacity decreases as the concentration of the solute increases.

Any inhomogeneity such as voids, cracks etc. increases the damping capacity because of the energy dissipation at the flaw-site. As far as the question of how small a defect can be detected using damping measurement technique is concerned, it is decided by the fact that whether the energy dissipation at the flaw site is a significant proportion of the total normal energy dissipation or not. A specimen without discontinuities has same value of damping in all directions whereas specimens having discontinuities have anistropy of damping. Quenching cracks and intrigranular corrosion also result in increased damping.

Damping measurement technique of dynamic testing have been used in past to evaluate the bond quality as well. However, acousto-ultrasonic technique provides a better solution to the problem of evaluating the strength of adhesive bonded joints.

8.11 SPECTROCHEMICAL ANALYSIS

Spectrochemical analysis is one of the commonest technique used for the analysis of various constituents of an alloy. Many commercial spectrochemical analysis instruments are available and they provide very quick and accurate analysis. There are several books on the subject of spectroscopy and a few of them are listed below for ready reference:

1. Agarwal, B.K., *X-ray Spectroscopy—An Introduction,* Springer-Verlag, Berlin, 1991.

2. Borde, W.R., *Chemical Spectroscopy*, John Wiley and Sons., Inc., New York, 1939.

3. Broekaert, J.A.C., *Analytical Atomic Spectrometry with Flames and Plasmas*, Wiley-VCH Verlag GmbH & Co., Weinheim, Germany, 2005.

4. Harrison, G.R., Lordd, R.C., and Lofbourow, J.R., *Practical Spectroscopy*, Prentice-Hall, Inc., Englewood Cliffs, NJ, 1949.

5. Ingle, J.D. and Stanley, R.C., *Spectrochemical Analysis*, Prentice-Hall, Englewood Cliffs, NJ, 1988.

6. Mirabella, F.M., *Modern Techniques in Applied Molecular Spectroscopy*, John Wiley & Sons, Inc., New York, 1998.

7. Muller, R.O., *Spectrochemical Analysis by X-ray Fluorescence*, Plenum Press, New York, 1972.

8. Sawyer, R.A., *Experimental Spectroscopy*, Prentice-Hall, Inc., Englewood Cliffs, NJ, 1944.

9. Twyman, F., *The Spectrochemical Analysis of Metals and Alloys*, Chemical Publishing Co., Brooklyn, NY, 1941.

10. Twyman, F., *Metal Spectroscopy*, Charles Griffin and Co. Ltd., London, 1951.

Spectrochemical analysis is a very accurate technique and sensitivity obtainable in parts per million. The basis of the spectrochemical analysis is that every chemical element produces a characteristic set of spectrum lines during spectrochemical analysis tests and therefore, these elements can easily be identified and quantified.

Any atom or molecule can be excited by the help of an arc, spark, flame or discharge tube. These excited atoms or molecules emit light of a particular wavelength, *i.e.*, every excited atom or molecule has its own characteristic wavelength light. When this light is passed through a prism or through a grating, dispersion of this light takes place. This dispersion of light results in production of a characteristic spectrum for each individual atom/molecule which is used to provide an identity to these atoms/molecules and also it becomes possible to quantify these atoms/molecules. Thus spectrochemical analysis instruments provide the percentage of different metallic and non-metallic constituents present in an alloy.

8.12 THICKNESS MEASUREMENT USING BETA GAUGE AND BETA BACKSCATTER GAUGE

Beta rays find extensive applications in the field of thickness measurement. Both absorption type beta gauge and backscatter type beta gauge are used for thickness measurement. These devices are very accurate and specially suitable for measuring thickness of extremely thin sheets, films, coating, plating, plastic coatings on wire etc. Both absorption type beta gauge and backscatter type beta gauge are schematically shown in Figs. 8.2 and 8.3 respectively.

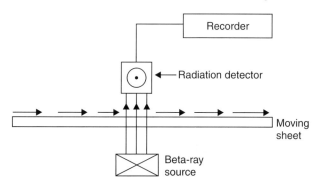

Fig. 8.2 Schematic arrangement of beta absorption gauge

Beta rays are nothing but high speed electrons which are emitted during disintegration of certain radioactive materials. Each beta ray emitter has a characteristic maximum energy and this maximum energy is expressed in MeV (mega electron volts).

While passing through any material, beta rays get impeded, *i.e.*, the high speed electrons are slowed down. This is due to the collisions of these high speed electrons with the electrons of the material through which it is passing. What thickness of a particular material will completely absorb the beta rays (*i.e.*, the range of beta rays of that particular material), depends upon the electron density of that particular material and upon the initial energy of the radiation. Using the following empirical relationship, one may compute the thickness which will completely absorb the beta rays having known maximum energy value.

$$t = 0.54\ E_m - 0.16\rho$$

where ρ = density of material (gm/cm^3) ,

   t = thickness of absorbing material (cm),

   E_m = maximum energy of beta rays in MeV.

Besides absorption, whenever beta rays impinge on a material, a portion of it gets reflected or backscattered. The intensity of backscattered radiation increases with increase in thickness of material on which the beta rays are impinging. This trend continues till the thickness equals approximately one-fifth of the range of beta rays for that particular material. A further increase in thickness does not increase the intensity of the backscattered radiation. The ability of a material to backscatter beta rays depends upon the atomic number (Z) or the material.

In beta-ray backscatter gauge (Fig. 8.3) the radiation detector is so placed that the direct radiation from the radiation source does not reach the detector. Some type of shield is also used for shielding the detector from direct radiation. Only the reflected or backscattered radiation is allowed to reach the detector. Also, the relative position of radiation source, radiation detector and the sample under test is so adjusted that one obtains maximum detector response.

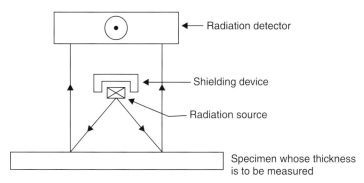

Fig. 8.3 Schematic arrangement of beta backscatter gauge

Using beta-ray backscatter gauge, one may very accurately measure thicknesses of extremely thin coatings, such as tin or zinc coatings on steel, barium coating on photographic paper, nickel or chromium platings, plastic coatings on wires etc.

Bibliography

1. *Acoustics of Wood,* Springer Series in Wood Science, Springer Verlag, Berlin, 2006.

2. *Applied Spectroscopy*, An International Journal from the Society of Applied Spectroscopy, Fredrick, MD.

3. *ASM Metals Handbook,* ASM Metals Handbook, Vol. 11, Non-Destructive Inspection and Quality Control, ASM, Materials Park, OH.

4. *ASTM Standards,* American Society for Testing of Metals, Philadelphia, PA.

5. Birnbaum, G. and Free, G. (Eds.), *Eddy-Current Characterization of Materials and Structures*, ASTM-STP 722, Philadelphia, 1981.

6. Blitz, J. and Simpson, G., *Ultrasonic Methods of Non-Destructive Testing*, Chapman & Hall, London, 1996.

7. Bray, D.E. and Stanley, R.K., *Non-Destructive Evaluation: A Tool in Design, Manufacturing and Service*, CRC Press, Boca Raton, Fl, 1997.

8. Cartz, L., *Non-Destructive Testing*, ASM International, Materials Park, OH, 1995.

9. Drouillard, T.F., *Acoustic Emissions—A Bibliography with Abstracts*, Plenum Press, NY, 1979.

10. Duke, J.C. Jr. (Ed.), *Acousto-Ultrasonics: Theory and Applications*, Plenum Press, New York, 1988.

11. *Experimental Mechanics*, An International Journal by Springer Verlag, Berlin.

12. *Experimental Techniques,* Journal of the Society for Experimental Mechanics, Published by Blackwell Publishing.

13. Grandt, A.F., *Fundamentals of Structural Integrity: Damage Tolerant Design and Non-Destructive Evaluation*, John Wiley & Sons, Hoboken, NJ, 2004.

14. Halmshaw, R., *Non-Destructive Testing*, Edward Arnold Publishers, London, 1991.

15. Handbook *of Radiographic Apparatus and Techniques*, International Institute of Welding, London.

16. *Handbook on the Ultrasonic Examination of Welds*, International Institute of Welding, London.

17. Hayward, G.P., *Introduction to Non-Destructive Testing*, American Society for Quality Control, MI, 1978.

18. Hellier, C.J., *Handbook of Non-Destructive Testing*, McGraw Hill, NY, 2001.

19. Hull, B. and John V., *Non-Destructive Testing*, Springer Verlag, NY, 1988.

20. Jiles, D.C., *Principles of Materials Evaluation*, CRC Press, Boca Raton, Fl, 2008.

21. *Journal of Materials Science*, An International Journal by Springer.

22. *Journal of Non-Destructive Evaluation*, An International Journal by Springer.

23. *Journal of Testing and Evaluation*, An International Journal by ASTM, Philadelphia, PA.

24. Krautkramer, J. and Krautkramer, H., *Ultrasonic Testing of Materials*, Springer Verlag, Berlin, 1990.

25. Lamble, J.H., *Principles and Practice of Non-Destructive Testing*, John Wiley & Sons, Inc., 1963.

26. Libby, H.L., *Introduction to Electromagnetic Non-Destructive Test Methods*, Wiley-Interscience, New York, 1971.

27. Marcus, R.K. and Broekaert J.A.C. (Ed.), *Glow Discharge Plasmas in Analytical Spectroscopy*, Wiley, Chichester, 2003.

28. *Materials Evaluation*, An International Journal by Springer.

29. *Materials and Structures*, An International Journal by Springer.

30. McGonnagle, W.J. (Ed.), *Proceedings of the Symposium on Physics and Non-Destructive Testing*, Dayton, OH, Sept./Oct. 1964, Gordon and Breach Science Publishers, NY, 1964.

31. McGonnagle, W.J., *Non-Destructive Testing Techniques*, Gordon and Breach Science Publishers, NY, 1964.

32. McGonnagle, W.J., *Non-Destructive Testing*, McGraw Hill Book Co., Inc., NY, 1961.

33. McMaster, R.C., *Non-Destructive Testing Handbook*, Vol. I and II, American Society for Metals, Columbus, Ohio, 1982.

34. *Metals Handbook*, ASM International, Materials Park, OH, 1998.

35. *NDT International*, An International Journal by Springer.

36. *NDT & E*, An International Journal by Elsevier Ltd.

37. Ness, S. and Sherlock, C.N., *Non-Destructive Testing Overview*, ASNT, Columbus, OH, 1996.

38. Nichols, R.W. (Ed.), *Acoustic Emission*, Applied Science Publishers Ltd., Barking, Essex, 1976.

39. *Physics of Non-Destructive Testing with Particular Reference to Some of the New Aspect*, British Journal of Applied Physics, Supplement No. 6, 1957.

40. Rokhlin, S.I., Datta, S.K. and Rajapakse, Y.D.S. (Eds.), *Ultrasonic Characterisation of Mechanics of Interfaces*, Applied Mechanics Division (AMD) of ASME, Vol. 177, NY, 1993.

41. Sachse, W., Roget, J. and Yamaguchi, K. (Eds.), *Acoustic Emission: Current Practice and Future Directions*, ASTM-STP 1077, Philadelphia, 1991.

42. Summerscales, J., *Non-Destructive Testing of Fibre Reinforced Plastics Composites*, Vol. 1 and 2, Elsevier Applied Science Publishers, London, 1987 and 1990.

43. Sharpe, R.S. (Ed.), *Research Techniques in Non-Destructive Testing,* Vol. 1 to 8, Academic Press, London, 1970 to 1985.

44. Thompson, D.O. and Chimentl, D.E., *Review of Progress in Quantitative Non-Destructive Evaluation*, American Institute of Physics, Melville, NY, 2008.

45. Vahaviolos, S.J. (Ed.), *Acoustic Emission: Standards and Technology Update*, ASTM-STP 1353, Philadelphia, 1999.

46. Welz, B. and Sperling, M., *Atomic Absorption Spectrometry*, Wiley VCH, Weinheim, 1999.

Index